Methods and Tools in Biosciences and Medicine

Microinjection,
edited by Juan Carlos Lacal, Rosario Perona and James Feramisco, 1999
DNA Profiling and DNA Fingerprinting,
edited by Jörg Epplen and Thomas Lubjuhn, 1999
Animal Toxins – Facts and Protocols,
edited by Hervé Rochat and Marie-France Martin-Eauclaire, 2000
Methods in Non-Aqueous Enzymology,
edited by Munishwar Nath Gupta, 2000
Techniques in Molecular Systematics and Evolution,
edited by Rob DeSalle, Gonzalo Giribet and Ward Wheeler, 2002
Methods for Affinity-Based Separations of Enzymes and Proteins,
edited by Munishwar Nath Gupta, 2002
Analytical Biotechnology,
edited by Thomas G.M. Schalkhammer, 2002
Prokaryotic Genomics,
edited by Michel Blot, 2003

Prokaryotic Genomics

Edited by
Michel Blot

Birkhäuser Verlag
Basel · Boston · Berlin

Editor
Prof. Dr. Michel Blot
Université Joseph Fourier
460 rue de la piscine
38400 St Martin d'Hères
France

Library of Congress Cataloging-in-Publication Data
Prokaryotic genomics / edited by Michel Blot.
     p. cm. – (Methods and tools in biosciences and medicine)
  Includes bibliographical references and index.
  ISBN 3764365978 (alk. paper) – ISBN 376436596X (soft cover : alk. paper)
    1. Bacterial genetics–Laboratory manuals. 2. Genomics–Laboratory manuals.
  3. Proteomics–Laboratory manuals. I. Blot, Michel. II. Series.

Deutsche Bibliothek Cataloging-in-Publication Data
Prokaryotic genomics / ed. by Michel Blot. – Basel;
Boston; Berlin: Birkhäuser, 2002
(Methods and tools in biosciences and medicine)
ISBN 3-7643-6597-8
ISBN 3-7643-6596-X

ISBN 3-7643-6597-8 Birkhäuser Verlag, Basel – Boston – Berlin
ISBN 3-7643-6596-X Birkhäuser Verlag, Basel – Boston – Berlin

In memory of Michel Blot

Michel Blot died at the age of 42 years in a tragic mountain accident on the 8th of September 2002. During his unusual and prematurely terminated career as a biologist he has conducted research in fields ranging from population biology to molecular genetics. Michel Blot's initial area of interest was classical population biology. After his theoretical training at the Pierre and Marie Curie university in Paris, he carried out part of his thesis work on the Kerguelen islands, in the extreme southern part of the Indian ocean, studying the population genetics of an indigenous mollusk. He received his Ph.D. from the university Paris VII in 1989. During his post-doctoral work in the laboratory of Werner Arber (laureate of the Nobel price of medicine in 1978) in Basel (Switzerland), he switched subject and became interested in bacterial genetics.

Michel Blot was appointed full professor at the University Joseph Fourier in Grenoble, France, in 1995 and set out to elucidate the molecular mechanisms of the evolution of bacterial cultures. Inspired by his background in population biology, he did not consider a bacterial culture as a population of millions of identical individuals but as cells that mutate, thereby producing a heterogeneous population that provides the substrate upon which evolutionary selection can act. During the past few years Michel and his group, with an ever-growing array of international collaborators, have established the role of insertion sequences as a motor of bacterial evolution. He cleverly exploited bacterial cultures that had been stored or grown for many years, making double usage of insertion sequences as molecular markers as well as causal elements of evolution. His research demonstrated how these elements provide genetic diversity and promote the evolutionary adaptation of bacteria to the environment in which they live. He has thus explored the entire spectrum from population biology to the study of molecular mechanisms of the regulation of the expression of genes involved in evolutionary adaptation.

The scientific merits of Michel Blot were recognized and honored by national and international science agencies and he was appointed junior member of the very prestigious Institut Universitaire de France in 1998. His success certainly rested partly on his extraordinary ability to convey to an audience, first-year students and established scientists alike, his enthusiasm and excitement for scientific research.

Michel Blot the scientist was well recognized, but Michel Blot the science politician has also left very distinctive marks in the national scientific community. Upon his arrival in Grenoble, he started off an initially very small microbiology laboratory and he invested a large amount of energy to make the laboratory grow and to provide an environment conducive to scientific research. He did succeed in creating a laboratory where the scientific imagination can thrive. His combative style made Michel some enemies as well as friends, who will retain a lasting respect, affection and gratitude.

Michel Blot will be most missed by his wife and his two sons, but his absence will also be felt by all those working in the fields of bacterial and molecular evolution. We have lost, far too soon, an exceptional scientist and colleague who had still much to give professionally and personally.

Contents

List of Contributors

AVRAMOVA, LARISA, Department of Biological Sciences, Purdue University, West Lafayette, IN 47907, USA; e-mail: la@bilbo.bio.purdue.edu

BOULOC, PHILIPPE, Laboratoire des Réseaux de Régulations, Institut de Génétique et Microbiologie, Université Paris-Sud, CNRS UMR8621, 91405 Orsay cedex, France; e-mail bouloc@infobiogen.fr

CAMACHO, EVA M., Departamento de Genetica, Universidad de Sevilla, Apartado 1095, 41080 Sevilla, Spain; e-mail: ecamacho@us.es

CASADESUS, JOSEP, Departamento de Genetica, Universidad de Sevilla, Apartado 1095, 41080 Sevilla, Spain; e-mail: casadesus@us.es

CHONG, SHAORONG, New England Biolabs, Inc., 32 Tozer Road, Beverly, MA 01915, USA; e-mail: chong@neb.com

DALE, COLIN, Department of Biochemistry, University of Arizona, Tucson, AZ 85721, USA; e-mail: cdale@email.arizona.edu

DATSENKO, KIRILL A., Department of Biological Sciences, Purdue University, West Lafayette, IN 47907, USA; e-mail: datsenko@bilbo.bio.purdue.edu

DENIZOT, FRANÇOIS, Laboratoire de Chimie Bactérienne, IBSM-CNRS, 31 Chemin Joseph Aiguier, 13402 Marseille cedex 20, France; e-mail: denizot@ibsm.cnrs-mrs.fr

DI LALLO, GUSTAVO, Dipartimento di Biologia, Università di Roma "Tor Vergata", Via della Ricerca Scientifica, 00133 Roma, Italy; e-mail: dilallo@uniroma2.it

D'ARI, RICHARD, Institut Jacques Monod, CNRS Universités Paris 6,7, 2 place Jussieu, 75251 Paris cedex 05, France; e-mail: dari@ijm.jussieu.fr

GHELARDINI, PATRIZIA, Centro Acidi Nucleici del CNR, c/o Dipartimento di Genetica e Biologia Molecolare, Università "La Sapienza", P.le Aldo Moro 5, 00185 Roma, Italy; e-mail: ghelardini@uniroma2.it

HOUSEWEART, CHAD, Genome Therapeutics Corporation, 100 Beaver Street, Waltham, MA 02453, USA

JOURLIN-CASTELLI, CÉCILE, Laboratoire de Chimie Bactérienne, IBSM-CNRS, 31 Chemin Joseph Aiguier, 13402 Marseille cedex 20, France; e-mail: jourlin@ibsm.cnrs-mrs.fr

KENNEY, TERESA J., Genome Therapeutics Corporation, 100 Beaver Street, Waltham, MA 02453, USA

KIM, SOO-KI, Department of Animal Products and Environmental Science, Konkuk University, 1 Hwayang-dong, Gwangin-gu, Seoul, 143–701, Korea; e-mail: mebong7@hanmail.net

LAUB, MICHAEL T., Bauer Center for Genomics Research, Harvard University, Cambridge, MA 02138, USA; e-mail: laub@cgr.harvard.edu

LELONG, CÉCILE, Plasticité et Expression des Génomes Microbiens, CNRS FRE2383, Université Joseph Fourier, 38041 Grenoble cedex, France; e-mail: cecile.lelong@ujf-grenoble.fr

MALOY, STANLEY, Center for Microbial Sciences, San Diego State University, 5500 Campanile Drive, San Diego, CA 92182–4614, USA; e-mail: smaloy@sciences.sdsu.edu

METZGAR, DAVID, Mail Code BCC-379, The Scripps Research Institute, 10550 North Torrey Pines Rd, La Jolla, CA, 92037, USA; e-mail: dmetzgar@hermes.scripps.edu

OCHMAN, HOWARD, Department of Biochemistry, 233 Life Sciences, Southern University of Arizona, Tucson, AZ 85721, USA; e-mail: hochman@email.arizona.edu

PAOLOZZI, LUCIANO, Dipartimento di Biologia, Università di Roma "Tor Vergata", Via della Ricerca Scientifica, 00133 Roma, Italy; e-mail: paolozzi@bio.uniroma2.it

PERLER, FRAN, New England Biolabs, 32 Tozer Road, Beverly, MA 01915, USA; e-mail: perler@neb.com

RABILLOUD, THIERRY, DRDC/BECP, CEA Grenoble, 17, Avenue des Martyrs, 38054 Grenoble cedex 9, France; e-mail: thierry@sanrafael.ceng.cea.fr

ROSENZWEIG, R. FRANK, Division of Biological Sciences, University of Montana, Missoula, MT 59812, USA; e-mail: rrose@selway.umt.edu

SCHNEIDER, DOMINIQUE, Plasticité et Expression des Génomes Microbiens, CNRS FRE2383, Université Joseph Fourier, 38041 Grenoble cedex, France; e-mail: dominique.schneider@ujf-grenoble.fr

SMITH, WENDY, Department of Ecology and Evolutionary Biology, University of Arizona, Tucson, AZ 85721, USA; e-mail: wasmith@email.arizona.edu

SWIFT, SIMON, Division of Molecular Medicine and Pathology, Faculty of Medical and Health Sciences, University of Auckland, Private Bag 92019, Auckland, New Zealand; e-mail: s.swift@auckland.ac.nz

THIERAUF, ANNE, Department of Microbiology, University of Illinois, 601 S. Goodwin Ave, Urbana, IL 61801, USA; e-mail: thierauf@students.uiuc.edu

TINLAND, BRUNO, Monsanto Europe Africa, Avenue de Tervuren 270–272, 1150 Brussels, Belgium; e-mail: bruno.tinland@monsanto.com

VINELLA, DANIEL, Institut Jacques Monod, CNRS-Universités Paris 6,7, 2 place Jussieu, 75251 Paris cedex 05, France; e-mail: vinella@ijm.jussieu.fr

WALKER, SCOTT, Schering-Plough Research Institute, 2015 Galloping Hill Road, 4700 Kenilworth, NJ 07033, USA; e-mail: scott.walker@spcorp.com

WANNER, BARRY L., Department of Biological Sciences, Purdue University, West Lafayette, IN 47907, USA; e-mail: BLW@bilbo.bio.purdue.edu

ZHOU, LU, Department of Biological Sciences, Purdue University, West Lafayette, IN 47907, USA; e-mail: zhou@purdue.edu

Preface

This manual reflects practical approaches to handling bacteria in the laboratory. It is designed to recall historical methods of bacterial genetics that have had recent developments and to present new techniques that allow full genome analysis. It has been written for microbiologists who need to group their protocols at the state of the art of a new millennium and also for scientists in other fields of life sciences who need to use bacteria for their research. Teachers, graduate students, and postdocs also will benefit from having these protocols to help them understand modern bacterial genetics.

I learned so much from these contributions from my colleagues that I have no doubt about the daily usefulness of this book.

April 2002   Michel Blot

Abbreviations

Acyl-HSL *N*-acyl homoserine lactone
Amp or Ap ampicillin
C carboxy
C10-HSL *N*-decanoyl-L-homoserine lactone
C12-HSL *N*-dodecanoyl-L-homoserine lactone
C14-HSL *N*-tetradecanoyl-L-homoserine lactone
C4-HSL *N*-butanoyl-L-homoserine lactone
C6-HSL *N*-hexanoyl-L-homoserine lactone
C8-HSL *N*-octanoyl-L-homoserine lactone
Cam or Cm chloramphenicol
CBD chitin binding domain
CHEF contour clamped homogenous electric field
CI consistency index
CRIM conditional-replication, integration, and modular
dCTP deoxycytidine triphosphate
deg. C degrees Celcius
DKP diketopiperazine
DMF dimethylformamide
DMSO dimethylsulfoxide
DNA deoxyribonucleic acid
dNTP deoxynucleotide dATP, dCTP, dGTP, dTTP
DTT dithiothreitol
EBU Evans Blue-Uranine
ECL enhanced chemiluminescence
ECOR *Escherichia coli* collection of reference
EDTA ethylamine diamine tetraacetic acid
EGTA ethylene glycol-*bis*(β-aminoethyl ether) N,N,N′,N′-tetraacetic acid
EPL expressed protein ligation
GC guanine/cytosine
GFP green fluorescent protein
HPLC high pressure liquid chromatography
HT high transducing
Int integrase
int integrase; mutation prevents formation of stable lysogens
IPL intein-mediated protein ligation
IPTG isopropyl β-D-thiogalactopyranoside
Kan Kanamycin
kb kilobase
kDa kiloDalton
LB Luria-Bertani broth
βME β-mercaptoethanol
MESNA 2-mercaptoethanesulfonic acid
min. minimum

moi multiplicity of infection
N amino
NMR nuclear magnetic resonance
3-OH-C14:1–HSL *N*-(3-hydroxy-7-cis-tetradecanoyl)homo-serine lactone
3-OH-C4-HSL *N*-3-hydroxybutanoyl-L-homoserine lactone
ONPG *o*-nitrophenyl β-D-galactopyranoside
ORF open reading frame
OTG 1-S-octyl-β-D-thioglucoside
3-oxo-C10-HSL *N*-3-oxodecanoyl-L-homo-serine lactone
3-oxo-C12-HSL *N*-3-oxododecanoyl-L-homoserine lactone
3-oxo-C14- HSL *N*-3-oxotetradecanoyl-L-homoserine lactone
3-oxo-C4-HSL *N*-3-oxobutanoyl-L-homoserine lactone
3-oxo-C6-HSL *N*-3-oxohexanoyl-L-homoserine lactone
3-oxo-C8-HSL *N*-3-oxooctanoyl-L-homoserine lactone
pac packaging site; DNA site where phage packaging is initiated
PCR Polymerase chain reaction
PFGE pulsed field gel electrophoresis
pfu plaque forming units
PMSF phenylmethylsulfonyl fluoride
QS quorum sensing
QSB quorum sensing blocker
RFLP restriction fragment length polymorphism
T0 time zero, initial cell population
T15, 30, 45 15, 30, 45 doublings of the initial population
Tc tetracycline
TCCP tris-(2-cyanoethyl)phosphine
TCEP tris-(2-carboxyethyl)phosphine
TE tris-EDTA buffer
TLC thin layer chromatography
T$_m$ melting temperature
Tris-HCl tris(hydroxymethyl)amino-methane-hydrochloride
TYE tryptone-yeast extract
v:v volume:volume
vir virulent
w:v weight:volume
X-gal 5-bromo-4-chloro-3-indolyl-(-D-galactoside)
Xis excisionase

Physical Analysis of Chromosome Size Variation

Colin Dale, Wendy Smith and Howard Ochman

Contents

1 Introduction

Early characterization of genetic material from a wide range of organisms involved the determination of base composition and genome size. Aside from the intrinsic value of such information, these properties were studied because they could be obtained for the large number of samples where cytogenetic and transmission genetic analysis was onerous or obscure. As it turned out, these general features divulged some of the most fundamental aspects of gene and genome organization and evolution. The base compositional differences among bacteria led to theories about mutational processes that foreshadowed the neutral theory of molecular evolution [1–3] and, among eukaryotes, to the discovery of the isochore structuring within chromosomes [4]. With respect to genome-size variation, the results were equally consequential. Across life forms, there seemed to be little relationship between the amount of genetic material and the degree of organismal complexity (the so-called "C-value paradox"), which has led to inquiries about the amounts, the accumulation, and the function of non-coding DNA in genomes [5–9]. Within bacteria, genome-size would appear to have direct consequences on the biology of an organism:

Methods and Tools in Biosciences and Medicine
Prokaryotic Genomics, ed. by M. Blot
© 2003 Birkhäuser Verlag Basel/Switzerland

because of the high coding content of bacterial DNA, variation in genome-size implies differences in the absolute number of genes.

The sizes of microbial genomes were assessed by thermal denaturation and/or reassociation kinetics [10–12], sedimentation and buoyant density [13, 14], and electron microscopic visualization [15]. However, the advent of pulsed-field gel electrophoresis (PFGE) [16, 17] clearly changed the way that chromosomes could be studied and the types of questions addressed. Moreover, the sizes of DNA fragments that could be readily resolved by PFGE were ideally suited to the known size range of bacterial chromosomes. Therefore, it was not surprising that the technique was rapidly adopted by geneticists, microbiologists, population biologists, epidemiologists, and taxonomists as a means to examine bacterial genomes. At last count, well over 200 bacterial species, and numerous samples within species, were characterized by PFGE [18]. The following sections describe the PFGE procedures used in our laboratory to examine genome-size variation and genetic polymorphism within enteric bacteria. These procedures can be readily adapted to investigate genome-size variation within any microbial group.

2   Methods

Variations in PFGE methodology have been developed to accommodate differences in bacterial growth rate, cellular composition, and genome size. The procedures detailed below are tailored to the preparation and analysis of genomic DNA from Escherichia coli using the contour clamped homogenous electric field (CHEF) electrophoresis method [19]. The CHEF method can be used to resolve DNA fragments of up to 10 Mbp in size and is currently the method of choice for most applications.

Protocol 1   Preparation of bacteria in agarose

1. Pellet cells from a 2 ml overnight bacterial culture (8000 × g, 1 min).
2. Wash bacterial pellet in TE (10 mM Tris·Cl [pH 8.0], 1 mM EDTA) and resuspend in 100 µl TES (50 mM Tris·Cl [pH 8.0], 100 mM EDTA, 25% [w/v] sucrose).
3. Add 20 µl 4 mg/ml lysozyme (in TES) and 180 µl 1% InCert agarose (FMC) in TES.
4. Transfer the bacteria/agarose suspension into 2-mm thick plastic molds and solidify at 4 °C.

Protocol 2 Preparation of bacterial DNA

1. Following solidification, transfer bacterial plugs into a minimum of 50 volumes of freshly prepared deproteinizing solution (0.5 M EDTA [pH 8.0], 1% [w/v] sarkosyl, 0.2 mg/ml proteinase K).
2. Allow digestion to proceed for 48 h at 37 °C, replacing the deproteinizing solution after 24 h.
3. Remove plugs from deproteinizing solution and transfer to 100 volumes of wash solution (50 mM EDTA, pH 8.0).
4. Wash plugs three times at room temperature with gentle agitation to remove proteinase and detergent prior to restriction digestion.
5. Optional: Wash plugs for 30 min in wash solution containing 10 µM PMSF (prepared fresh from a stock of 10 mM PMSF in ethanol, stored in a light-proof bottle at –20 °C). PMSF is a proteinase inhibitor and its use can improve subsequent digestion efficiency. This chemical is toxic and should be handled with great care. Eliminating the PMSF step still generates consistent and reliable results.
6. Plugs can be stored at 4 °C in 50 mM EDTA (pH 8.0) for several months.

Protocol 3 Restriction digestion of bacterial DNA

1. After washing and storage, plugs should be trimmed to a size convienient for restriction digestion in 1.5 ml Eppendorf tubes. To ensure adequate digestion, we recommend a size no greater than 2 mm × 5 mm × 5 mm.
2. The cut plugs should be equilibrated in 100 volumes of TE buffer at room temperature for 2 h and then equilibrated overnight (16 h) in 10 volumes of 1 × restriction buffer.
3. Prior to digestion, the restriction buffer should be removed and replaced. We conducted restriction digestions overnight in 1 × restriction buffer in a 250 µl reaction containing 0.1 units of restriction enzyme/microlitre. Many enzyme manufacturers provide specific recommendations for reaction volume, enzyme concentrations, and reaction conditions.
4. Following restriction digestion, electrophoresis should be carried out as soon as possible to avoid degradation of the plugs.

Protocol 4 Pulsed-field gel electrophoresis

Separation of DNA fragments during CHEF PFGE is governed by electric field strength, pulse time, temperature, buffer ionic strength, and gel concentration. In practice, electrophoresis can be optimized by the appropriate selection of gel concentration, run time, field strength, and pulse parameters. Despite several analytical methods designed to obtain optimal resolution, trial and error is sometimes needed to achieve an acceptable degree of separation. Occasionally, it is necessary to run the same plugs under different conditions in order to resolve all of the DNA fragments generated by restriction digestion. See Table 1 for some suggested conditions suitable for resolving DNA fragments in different size ranges.

Table 1   Recommended PFGE parameters

Size range	Initial switch time[1]	Final switch time[1]	Run time (h)
1–25 Kbp	0.1 s	2 s	10
5–50 Kbp	2 s	10 s	18
50–300 Kbp	7 s	25 s	30
100–450 Kbp	12 s	40 s	34
200–900 Kbp	25 s	75 s	40
400–1500 Kbp	50 s	120 s	60

[1]   Utilizing a linear switching ramp, a 1% agarose gel at 14 °C

1. Prepare 4 l electrophoresis buffer (0.5 × TBE) and use some of this buffer to prepare a gel containing pulsed-field certified agarose.
2. Fill CHEF electrophoresis tank with remaining electrophoresis buffer and switch on buffer circulator and cooler (14 °C)
3. When the gel has completely solidified, insert digested plugs and marker plugs into available slots and seal with a drop of molten 1% agarose in 0.5 × TBE. We routinely use markers from NEB (Beverly, MA), including a yeast chromosome marker (225 Kbp to 1.9 Mbp) and a λ concatamer ladder (48.5 Kbp to 1 Mbp) when resolving bacterial genome fragment sizes.
4. When electrophoresis buffer has cooled to the appropriate temperature, transfer and secure gel in the PFGE apparatus, re-check all parameters, and start the run.
5. After the run, stain the gel with ethidium bromide and destain in water for at least 1 h. Prolonged destaining (up to 48 h) can improve visualization but may cause diffusion of smaller bands (< 100 Kbp).

Figure 1 shows the differences among strains of *E. coli* that have been identified by digestion with an 8-base cutter (*Not*I), followed by PFGE. As well as establishing and enabling comparisons of chromsome size of each of the strains, restriction fragment length polymorphisms (RFLPs) also are observed among strains. Bands that appear very brightly stained are apt to consist of co-

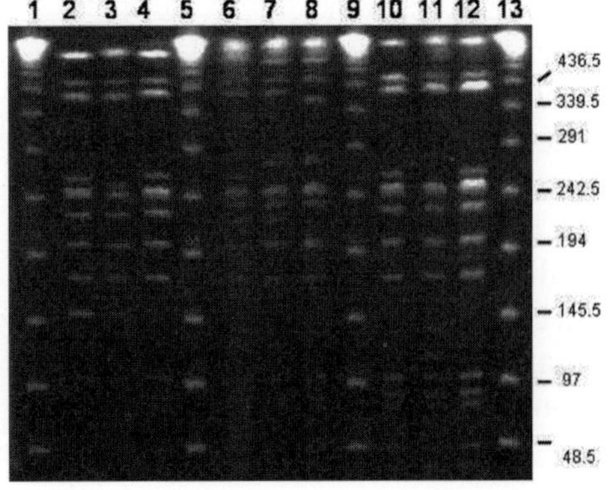

Figure 1   PFGE analysis of *E. coli* chromosomal DNA.
*Not*I-restricted genomic DNA from nine isolates of *E. coli* resolved by pulsed-field gel electrophoresis. Lanes 1, 5, 9, 13; Lambda concatamer PFGE ladder. All other lanes; *Not*I-restricted genomic DNA from isolates of *E.coli*. DNA fragments resolved by PFGE through a 1% gel (50 h, 170 V, 10–40 s switch time). Sizes of molecular weight standards shown.

migrating DNA fragments. To determine whether this is indeed the case, such bands are excised and redigested as described below.

Protocol 5   Resolving multiple fragments following electrophoresis

Occasionally, restriction digestion can generate DNA fragments that co-migrate during PFGE. For accurate determination of genome size, it is important to be able to resolve such multimers. This can be achieved by several different methods. First, it may be possible to separate co-migrating bands by optimizing electrophoretic conditions such that the resolution of fragments is enhanced over the desired size range (e. g. < 20 Kbp, 20–100 Kbp, and 100–700 Kbp). Also, conventional agarose gel electrophoresis should be applied for the separation of fragments < 20 Kbp. In the event that such optimization is insufficient, we have found that redigestion of the DNA fragment provides an accurate empirical means of analyzing mutiple DNA bands. In this method, the fragment of interest is physically excised from the gel and redigested with an alternate restriction enzyme. Resulting fragments generated by the second digestion reaction are resolved on a subsequent gel.

1. Excise the band of interest from gel in as small a piece of agarose as possible with a scalpel or razor blade.
2. Wash the gel band in 100 volumes of TE for 18 h to remove boric acid and excess EDTA. Change TE buffer at least twice during the washing period.
3. Equilibrate the gel band overnight in 10 volumes of 1 × restriction buffer and replace the restriction buffer prior to digestion.
4. Set up an overnight (16 h) restriction digest of the gel band in a 250 μl reaction containing 25–40 units of restriction enzyme. Be sure to check the manufacturer recommendations relating to reaction volume, enzyme concentrations, and reaction conditions.
5. Following restriction digestion, carry out PFGE as describe in protocol 4 to determine the complexity of the gel band.
6. Note that an apparently incomplete digestion may represent a case where one or more DNA species were not digested by a particular enzyme. In these cases, it is advantageous to digest bands with a number of different restriction enzymes.

2.1 Other considerations

Selecting appropriate restriction enzymes
For accurate resolution of genome sizes through PFGE, it is important to select restriction enzymes that cut a given DNA species at a low frequency (5–40 sites). For analysis of bacterial chromosomes, we favor the use of enzymes with 8-base recognition sites for DNA molecules > 3 Mbp and the use of non-degenerate 6-base cutters for molecules < 3 Mbp. If genome-size is unknown, we recommend the initial use of 8-base cutters.

Base composition (mol% G+C) dramatically affects the frequency of different restriction sites in any given genome. Fairly accurate estimates of base composition can be derived from a small amount of coding sequence data (not rDNA sequence data). When base composition is known, the frequencies of restriction enzyme cut sites can be estimated by probability theory. For example, in a host genome that is 60% G+C, the probability of finding G at any given site, p(G), is 0.3. The probability of finding a 6-base long 100% G+C restriction site (e. g., GGCGCC) is $(0.3)^6$, or 7.3×10^{-4}, or once in every 1.37 Kbp. In a genome composed of only 30% G+C, the probability of finding the same site is only 1.14×10^{-5}, or once in every 87.7 Kbp.

The distribution of restriction enzyme sites in a given genome also is affected by polynucleotide conservation in bacterial genomes [20]. For example, CTAG is known to be a rare tetranucleotide in bacterial genomes of $\geq$ 45% G+C, spanning an infrequently utilized leucine codon (CTA) or the rare (amber) stop codon (TAG). Similarly, the CC and GG dinucleotides are rarer than the GC dinucleotide and, thus, CCG and CGG are rare trinucleotides in bacterial genomes of $\leq$ 45% G+C. To reduce the frequency of cutting in a given bacterial genome, we recommend the selection of enzymes that incorporate these rare trinucleotides in their recognition sequences.

To demonstrate the effect of chromosome size, G+C content, and polynucleotide abundance on the frequency of restriction site availability, we analyzed bacterial genome sequence data available at the NCBI web site (http://www.ncbi.nlm.nih.gov). The results of this analysis are presented in Table 2.

Use of the I-CeuI intron encoded endonuclease
An extreme example of polynucleotide conservation is the presence of the I-*Ceu*I recognition site in bacteria genomes. The I-*Ceu*I recognition site (TAA CTA TAA CGG TCC TAA GGT AGC GA) is conserved within the rDNA operons of bacteria. Given that bacteria have between 1 and 15 copies of the rDNA operon within their genome [21], I-*Ceu*I is a good enzyme choice for preliminary genome-size determination. We recommend digesting DNA with very small amounts of the I-*Ceu*I enzyme (0.01 units/μl for a 4 h digestion), since reactions with I-*Ceu*I are prone to overdigestion.

3   Results and discussion

3.1 Accurate determination of bacterial genome-size by PFGE

Prior to the implementation of the large-scale bacterial genome sequencing projects, PFGE represented the only widely available method suitable for the accurate determination of bacterial genome size. Principally, the suitability of

Table 2 Number of rare-cutting restriction sites in selected bacterial chromosomes

Organism	Chromo-some Size (Kbp)	G+C (%)	Number of restriction sites in chromosome[1]							
			*Pac*I TTAATTAA	*Pme*I GTTTAAAC	*Sgf*I GCGATCGC	*Not*I GCGGCCGC	*Xba*I TCTAGA	*Spe*I ACTAGT	*Sma*I CCCGGG	*Sac*II CCGCGG
Buchnera sp. APS	640	26.3	184	12	1	0	170	129	8	9
Mycoplasma pneumoniae	816	40.0	173	62	11	2	87	282	77	68
Treponema pallidum	1138	52.3	7	11	40	20	96	40	160	295
Clostridium perfringens	3031	28.6	473	48	12	1	571	484	40	25
Synechocystis sp. PCC6803	3573	40.0	173	62	11	2	87	282	77	68
Caulobacter crescentus	4016	67.2	3	3	1003	884	116	29	1333	1452
Nostoc sp. PCC7120	6413	41.3	504	25	143	16	1642	29	1333	1452
Pseudomonas aeruginosa	6264	66.6	6	1	1470	979	99	37	2240	5878

[1] Rare tri- and tetra-nucleotides are underlined in the enzyme recognition sequences

this technique results from the fact that bacterial genome sizes fall within a range that can conveniently be analyzed by PFGE. At present, the sizes of the completely sequenced bacterial genomes range from 0.58 Mbp (*Mycoplasma genitalium*) to 8.7 Mbp (*Streptomyces avermitilis*), with a mean of around 3 Mbp.

The successful application of PFGE for genome-size estimation depends upon the availability of suitable rare-cutting restriction enzymes. Fortunately, a large variety of restriction enzymes are commercially available and a range of rare-cutting enzymes exists for every possible combination of bacterial genome-size and base composition. Table 2 illustrates how recognition site complexity affects the frequency of site availability for different enzymes in genomes of different sizes and base compositions. In general, most enzymes with 6-base recognition sites are useful only for the digestion of very small genomes (< 1Mbp) in which most 8-base cutters would cut very infrequently or not at all. For bacteria with genomes > 1 Mbp, the 8-base cutters are the enzymes of choice because they tend to generate a more acceptable number of resolvable restriction fragments. Certain bacterial species have significant base compositional bias, and this information can assist in the selection of rare-cutting restriction enzymes. If the genome is A+T rich, it makes sense to select enzymes with G+C rich recognition sites and *vice versa*.

For any given restriction enzyme digestion, the likelihood of generating two or more restriction fragments of similar size increases with the number of restriction enzyme cut sites in the template. Because similarly sized bands co-migrate during electrophoresis, these bands are often indistinguishable on pulsed-field gels. This problem can be addressed most easily by evaluating genomes with different restriction enzymes and comparing the total lengths of fragments generated by each enzyme. Multiple co-migrating fragments can be detected by visualization: ethidium bromide binds stoichiometrically and the intensity of staining is directly proportional to the amount of DNA. However, as genome-size increases, it becomes more difficult to identify restriction enzymes that cut infrequently. Thus, larger genomes are best analyzed initially with I-*Ceu*I digestion.

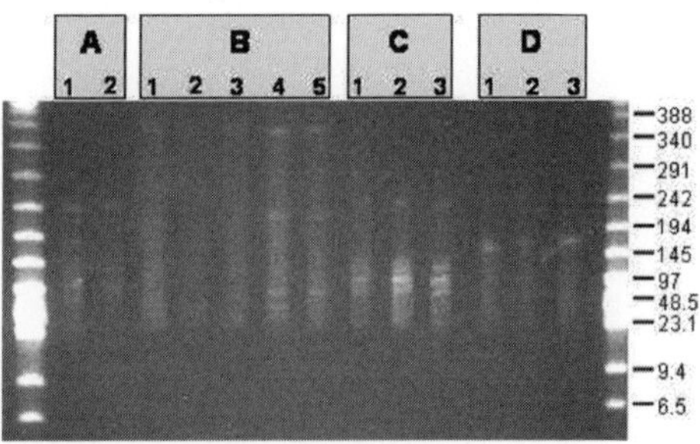

Figure 2   Detecting co-migrating DNA fragments by excision and redigestion. *Not*I-restricted DNA fragments from three isolates of *E. coli* excised from a pulsed-field gel and redigested with *Xba*I. Panels represent fragments of different sizes (**A**, 240 Kbp; **B**, 360 Kbp; **C**, 240 Kbp; **D**, 170 Kbp), and numbers (1–5) represent different isolates of *E. coli*. Fragments resolved by PFGE through a 1% gel (22 h, 200 V, 5–40 s switch time). Sizes of molecular weight standards shown.

In this chapter we also describe and demonstrate a robust empirical method useful for analyzing multiple bands (Fig. 2), whereby fragments are excised from the gel, digested with a second restriction enzyme, and separated by PFGE. The molecular weights of the resulting fragments can then be tallied to calculate how many multiples of the original fragment were present. Provided that suitable restriction enzymes are selected, this technique offers a fast and reliable means for resolving co-migrating DNA fragments.

References

1 Sueoka N (1961) Variation and heterogeneity of base composition of deoxyribonucleic acids: a compilation of old and new data. *J Molec Biol* 3: 31–40

2 Sueoka N (1962) On the genetic basis of variation and heterogeneity of DNA base composition. *Proc Natl Acad Sci USA* 48: 582–591

3 Freese E (1962) On evolution of base composition of DNA. *J Theor Biol* 3: 82

4 Bernardi G (2000) The compositional evolution of vertebrate genomes. *Gene* 259: 31–43

5 Britten RJ, Koehn DE (1968) Repeated sequences in DNA. *Science* 161: 529–540

6 Orgel LE and Crick FH (1980) Selfish DNA: the ultimate parasite. *Nature* 284: 604–607

7 Doolittle WF, Sapienza C (1980) Selfish genes, the phenotype paradigm and genome evolution. *Nature.* 284: 601–603

8 Cavalier-Smith T (1985) *The evolution of genome size.* John Wiley, New York

9 Holmquist GP (1989) Evolution of chromosome bands: molecular ecology of noncoding DNA. *J Molec Evol* 28: 469–486

10 Marmur J, Doty P (1962) Determination of the base composition of deoxyribo-nucleic acid from its thermal denaturation temperature. *J Molec Biol* 5: 109–118

11 Christiansen C, Bak AL, Sterderup A (1971) Genome-size determination of microbial DNA by renaturation: methodological considerations. *J Gen Microbiol* 64: xii

12 Brenner DJ, Fanning GR, Skerman FJ, Falkow S (1972) Polynucleotide sequence divergence among strains of *Escherichia coli* and closely related organisms. *J Bacteriol* 109: 933–965

13 Meselson M, Stahl F, Vinograd J (1957) Equilibrium sedimentation of macromolecules in density gradients. *Proc Natl Acad Sci USA* 43: 581–588

14 Schildkraut CL, Marmur J, Doty P (1962) Determination of base composition of deoxyribonucleic acid from its buoyant density in CsCl. *J Molec Biol* 4: 430–443

15 Cairns J (1963) The chromosome of *Escherichia coli. Cold Spring Harbor Symp Quant Biol* 28: 43–46

16 Schwartz DC, Cantor CR (1984) Separation of yeast chromosome-sized DNAs by pulsed field gradient gel electrophoresis. *Cell* 37: 67–75

17 Carle GF, Olson MV (1984) Separation of chromosomal DNA molecules from yeast by orthogonal-field-alternation gel electrophoresis. *Nucl Acids Res* 12: 5647–5664

18 Casjens S (1998) The diverse and dynamic structure of bacterial genomes. *Annu Rev Genet* 32: 339–377

19 Chu G, Vollrath D, Davis RW (1986) Separation of large DNA molecules by contour-clamped homogeneous electric fields. *Science* 234: 1582–1585

20 McClelland M, Jones R, Patel Y, Nelson M (1987) Restriction endonucleases for pulsed field mapping of bacterial genomes. *Nucl Acids Res* 15: 5985–6005

21 Klappenbach JA, Saxman PR, Cole JR, Schmidt TM (2001) rrndb: The Ribosomal RNA Operon Copy Number Database. *Nucl Acids Res* 29: 181–184

2 Genetic Mapping in *Salmonella enterica*

Josep Casadesus and Eva M. Camacho

Contents

1 Introduction

Genetic mapping has been a major task of classical genetics. Ascribing mutations to genes, and genes to chromosomes, has required the development of a large number of tools, both experimental and analytical, that lie at the core of genetical thinking. With the advent of alternative (and often simpler) molecular methods, some such tools have become obsolete, at least in practice. Others, however, survive. This chapter describes two methods for genetic mapping currently used in the gram-negative bacterium *Salmonella enterica*, a close

Methods and Tools in Biosciences and Medicine
Prokaryotic Genomics, ed. by M. Blot

relative of *Escherichia coli* that remains a model organism in both bacterial genetics and microbial pathogenesis.

The genetics of *Salmonella* was born in the 1950s, when Norton Zinder and Joshua Lederberg described generalized transduction of host DNA mediated by bacteriophage P22 [1]. P22 transduction–reviewed by S. R. Maloy in Chapter 6– has been exploited by a number of investigators, led by John Roth's group at the University of Utah, to design sophisticated procedures of genetic analysis. The two procedures described below–rapid mapping with locked-in Mu*d*-P22 hybrid prophages and genetic mapping by duplication segregation– make use of phage P22 and were described in the last decade. Neither method is directly applicable to bacterial species other than P22-sensitive Salmonellae, but the underlying principles might inspire similar procedures in other bacterial species.

1.1 Genetic mapping in *Salmonella*

Genetic mapping in *Salmonella* traditionally has used two approaches: conjugal transfer of DNA *via* Hfr formation [2] and cotransduction of nearby markers by "high-transducing" (HT) derivatives of bacteriophage P22 [3]. Cotransduction with P22 HT is probably the easiest and most reliable method of genetic mapping ever described [4], but it can be applied only to markers separated by less than 30 kb [3]. For several decades, large-scale mapping has relied on conjugal transfer. Many *Salmonella* strains contain F-like plasmids, some of which are self-transmissible [5]. However, the conjugation systems of *Salmonella* plasmids are tightly repressed, and conjugation occurs at very low frequencies [6]. In addition, most such plasmids are inefficient for chromosome mobilization. These difficulties can be overcome by using the *E. coli* F episome [2]. F can be transferred to *Salmonella*, and Hfr strains can be obtained upon F integration into the chromosome [2]. However, the absence of insertion elements IS*2* and IS*3* from the *Salmonella* chromosome [7] reduces the efficiency of Hfr formation. In the 1970s, the development of transposon technology introduced a number of refinements for Hfr formation in *Salmonella*, such as the use of transposable elements as "portable" regions of homology to direct F integration by homologous recombination [8, 9]. Despite this refinement, Hfr mapping in *Salmonella* still encounters problems derived from the instability of the Hfr donors.

Rapid mapping with locked-in Mu*d-P22 prophages*
A mapping procedure that can substitute for Hfr transfer in *S. typhimurium* was developed by Nick Benson and Barry Goldman a decade ago [10]. The method, originally devised by Phil Youderian, employs hybrids between two temperate bacteriophages, *Salmonella* phage P22 and coliphage Mu [11]. The Mu*d*-P22 hybrids constructed combine the P22 DNA packaging system with the transposition properties of phage Mu; as a consequence, the hybrids are able to

transpose randomly to the host chromosome. Each hybrid consists of about two-thirds of the phage P22 genome flanked by the ends of bacteriophage Mu; the construction also carries a chloramphenicol resistance marker [11].

Because Mu*d*-P22 hybrid prophages contain the P22 immunity region, they can be induced by DNA-damaging treatments. Upon induction, the entire prophage genome is replicated *in situ* but cannot excise because the construction lacks the P22 site-specific recombination region [10]. As a consequence, replication forks initiated at the P22 replication origin invade neighboring host DNA, causing a selective amplification of the chromosomal regions that flank the prophage. Replicated regions are then packaged into P22 capsids by the "headful" mechanism characteristic of phage P22 [12]. The first headful will package a portion of the defective hybrid prophage and some adjacent host DNA; further headfuls will consist of chromosomal DNA only [10]. Hence, induction of locked-in prophages leads to transduction of adjacent markers at high frequencies. Marker enrichment ranges between 50 and 1500 fold [10].

Packaging from P22 *pac* sites proceeds in an oriented manner [12]; thus, a given locked-in prophage can be expected to package flanking host DNA from only one side (Fig. 1). The packaging direction will depend on the orientation of the prophage itself. Two different Mu*d*-P22 prophages (Mu*d*P and Mu*d*Q) have been constructed, each packaging in one orientation [11]. When inserted at the same locus, Mu*d*P will package in one direction and Mu*d*Q in the other. The packaging capacity of a P22 capsid is about 43 kb, i. e., is, approximately one map minute [12]. The length of the amplified region allows packaging of 3–4 headfuls; however, a gradient of packaging efficiency is usually observed [10].

A collection of Mu*d*P and Mu*d*Q lysogens, each containing a different insertion, was constructed by Benson and Goldman [10]. The insertions are scattered along the *Salmonella* chromosome (Tab. 1). Transduction with lysates from this

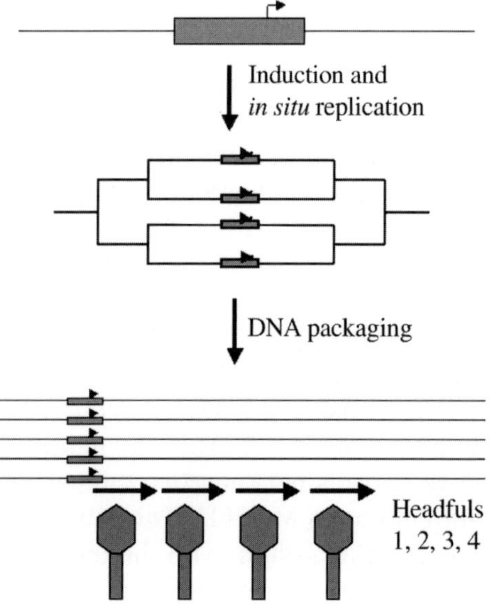

Figure 1 Induction of a locked-in Mu*d*-P22 prophage triggers *in situ* replication, which extends to chromosomal regions at both sides of the prophage. Induction also leads to synthesis and assembly of capsid components. Packaging of P22 DNA starts at the P22 *pac* site (indicated as an arrow) and proceeds unidirectionally to complete 3–4 headfuls, each 43 kb.

collection allows specialized transduction and mapping with a resolution of 1–4 minutes, provided that a selection procedure is available. The technique is directly applicable to any mutation for which the wild-type allele is selectable. In addition, Tn*10* insertions can be mapped by selecting loss of tetracycline resistance on appropriate media [13, 14]. Hence, a general procedure to map any mutation is the isolation of a linked Tn*10* insertion [15], which is then mapped by the Benson and Goldman procedure.

Table 1 A collection of lysogens carying Mu*d*-P22 prophages[a]

Strain	Genotype	Map location (centisome)
TT15222	*leuA414* r⁻ m⁺ *fels2 thr-458*::Mu*d*Q	0/100
TT15223	*leuA414* r⁻ m⁺ *fels2 thr-469*::Mu*d*P	0/100
TT15227	*leuA414* r⁻ m⁺ *fels2 nadC218*::Mu*d*Q	3
TT15228	*leuA414* r⁻ m⁺ *fels2 nadC220*::Mu*d*P	3
TT15229	*leuA414* r⁻ m⁺ *fels2 proA692*::Mu*d*Q	7
TT15233	*leuA414* r⁻ m⁺ *fels2 purE2155*::Mu*d*P	12
TT15234	*leuA414* r⁻ m⁺ *fels2 purE2155*::Mu*d*Q	12
TT15238	*leuA414* r⁻ m⁺ *fels2 nadA219*::Mu*d*P	17
TT15240	*leuA414* r⁻ m⁺ *fels2 putA1019*::Mu*d*P	25
TT15243	*leuA414* r⁻ m⁺ *fels2 aroD561*::Mu*d*Q	30
TT15244	*leuA414* r⁻ m⁺ *fels2 aroD561*::Mu*d*P	30
TT15245	*leuA414* r⁻ m⁺ *fels2 pyrF2690*::Mu*d*Q	37
TT15249	*leuA414* r⁻ m⁺ *fels2 zea-3666*::Mu*d*Q	40
TT15250	*leuA414* r⁻ m⁺ *fels2 zea-3666*::Mu*d*P	40
TT15251	*leuA414* r⁻ m⁺ *fels2 hisH9950*::Mu*d*P	44
TT15252	*leuA414* r⁻ m⁺ *fels2 hisH9950*::Mu*d*Q	44
TT15253	*leuA414* r⁻ m⁺ *fels2 cysA1586*::Mu*d*Q	53
TT15254	*leuA414* r⁻ m⁺ *fels2 cysA1586*::Mu*d*P	53
TT15255	*leuA414* r⁻ m⁺ *fels2 guaAB5641*::Mu*d*P	54
TT15259	*leuA414* r⁻ m⁺ *fels2 nadB226*::Mu*d*P	57
TT15260	*leuA414* r⁻ m⁺ *fels2 nadB226*::Mu*d*Q	57
TT16706	*leuA414* r⁻ m⁺ *fels2 zgc-3715*::Mu*d*P	62
TT16707	*leuA414* r⁻ m⁺ *fels2 zgc-3715*::Mu*d*Q	62
TT15262	*leuA414* r⁻ m⁺ *fels2 cysHIJ1574*::Mu*d*Q	64
TT17190	*leuA414* r⁻ m⁺ *fels2 zgi-3717*::Mu*d*P	68
TT17191	*leuA414* r⁻ m⁺ *fels2 zgi-3717*::Mu*d*Q	68
TT15264	*euA414* r⁻ m⁺ *fels2 cysG1573*::Mu*d*P	75
TT15265	*leuA414* r⁻ m⁺ *fels2 envZ1005*::Mu*d*P	76
TT17165	*leuA414* r⁻ m⁺ *fels2 envZ1005*::Mu*d*Q	76
SV2513	*leuA414* r⁻ m⁺ *fels2 xyl-561*::Mu*d*Q	80
TT15266	*leuA414* r⁻ m⁺ *fels2 pyrE2419*::Mu*d*Q	81
TT15267	*leuA414* r⁻ m⁺ *fels2 pyrE2419*::Mu*d*P	81
TT15269	*leuA414* r⁻ m⁺ *fels2 ilvA2648*::Mu*d*P	85

Strain	Genotype	Map location (centisome)
TT15274	*leuA414* r⁻ m⁺ *fels2 purD1874*::Mud*P*	90
TT15276	*leuA414* r⁻ m⁺ *fels2 melAB396*::Mud*P*	93
TT15277	*euA414* r⁻ m⁺ *fels2 purA1881*::Mud*P*	95
TT15638	*euA414* r⁻ m⁺ *fels2 zjh-3725*::Mud*P*	97
TT15638	*euA414* r⁻ m⁺ *fels2 zjh-3725*::Mud*Q*	97

[a]   All the strains were constructed by N. R. Benson and B. Goldman ([10] and unpublished data), except SV2513 [20].

Mapping can be carried out on a single Petri plate (or a few plates). For mapping, the strain carrying the mutation to be mapped is transduced with a collection of Mud-P22 lysates obtained by lysogen induction. Transfer of the selected allele gives a confluent spot of transductants, while the remaining spots appear similar to the background of cells that did not receive any phage [10]. Since both the position and the packaging direction of the donor prophage are known, the approximate position of the transduced marker can be easily inferred.

Lysates obtained upon induction of Mud-P22 hybrid prophages are made of P22 heads only [11]; hence, formation of complete phage capsids requires addition of P22 tails and head-tail assembly (see 3.1, page 17).

Genetic mapping by duplication segregation

Analysis of duplication segregation provides another method to localize a genetic marker in a chromosomal region. The procedure is based on the rationale that a marker introduced into a preexisting duplication will segregate together with the duplication, while a marker outside the duplication will not [16]. The marker to be mapped is transferred by P22 HT transduction into a recipient carrying a duplication with known endpoints [16]. The resulting isolates contain the duplication and the marker to be mapped. These isolates are then allowed to segregate (simply by growth under conditions that do not select the duplication), and haploid segregants are scored for the presence of the marker. Interpretation of the results is simple and straightforward: if all the segregants contain the marker, the latter maps outside the duplication interval. If the marker is found in only a fraction of the segregants, it maps inside the duplicated region (Fig. 2).

Construction of duplications with predetermined endpoints is achieved by an elegant method devised by Kelly Hughes and John Roth [17]. Two genetic elements properly positioned (e. g., two transposon insertions in the same orientation) provide homologies to generate a duplication with endpoints at the sites of the original insertions. Formation of the duplication requires a triple crossover; transductants generated by this rare event can be selected if the transposon insertions used cause auxotrophy. Selection of the transposon-borne antibiotic resistance on minimal plates yields prototrophic transductants carrying the desired duplication [17].

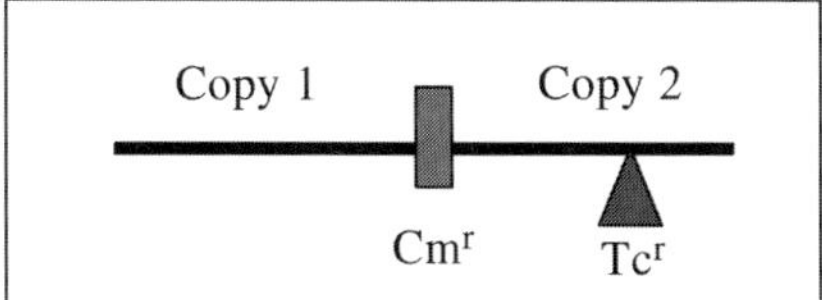

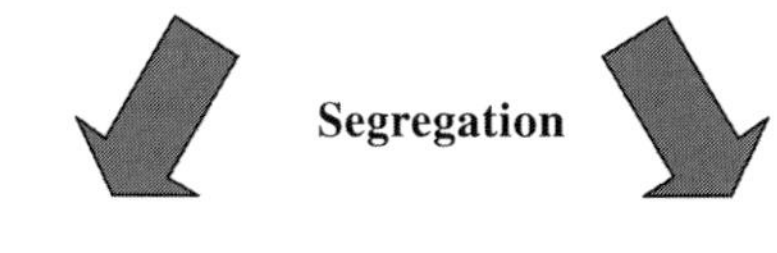

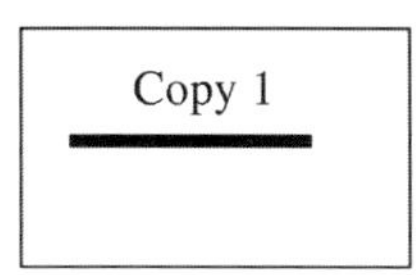

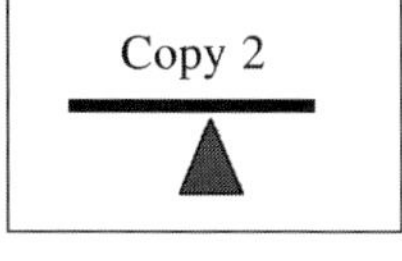

Figure 2 Segregation of a merodiploid carrying a chromosomal duplication held by a MudP/Q element (Cm^r) and a Tn10 (Tc^r) inserted in one copy of the duplicated region.

Maintenance of a transposon-held duplication requires continuous selection of the corresponding antibiotic resistance marker; however, strains carrying one such duplication can be frozen and later recovered on antibiotic medium. An alternative preservation procedure is the storage of P22 HT lysates grown on the merodiploid collection. If a haploid strain is transduced with one such lysate and the transposon-borne antibiotic resistance that holds the duplication is selected, the duplication will be reconstructed in the recipient [16, 17]. High-titer P22 HT lysates sterilized with chloroform are stable for several years if kept at 4 °C.

Table 2 A collection of *S. enterica* strains carrying chromosomal duplications with known endpoints [a]

Strain	Genotype	Duplication span (centisomes)
SV1604	Dup [*thr-469* *MudP*proA692*]	0–7
SV1603	Dup [*proA692* *MudQ*purE2164*]	7–12
SV1611	Dup [*purE2514* *MudP*purB1879*]	12–27
SV4200	Dup [*trp-248* *MudP*purA1881*]	38–44
SV3193	Dup [*hisH9962* *MudP*cysA1586*]	44–53
SV4015	Dup [*cysA1586* *MudP*purG2149*]	53–56
SV4193	Dup [*purG2149*MudQ*argA9001*]	56–64
SV4194	Dup [*argA9000*MudP*cysG1573*]	64–75
SV1601	Dup [*cysG1573* *MudQ*ilvA2642*]	75–85
SV4195	Dup [*ilvA2648*MudP*purA1881*]	85–95
SV4142	Dup [*purA1881*MudQ*thr-469*]	95–0

[a] Described in [16].

A collection of chromosomal duplications held by Mu*d*P/Q elements was recently constructed [16]. This collection permits mapping of a dominant mutation to any chromosomal interval except the terminus region, where duplications are not viable [16]. Each of the 11 strains listed in Table 2 is a merodiploid carrying a duplication held by a Mu*d*P/Q element. If maintained under selection, these strains are stable; however, segregation can be easily achieved by growth under conditions that do not select the duplication (e. g., in the absence of chloramphenicol). Mapping is carried out in three steps. First, the marker to be mapped is transferred by P22 HT transduction to all members of the 11-strain collection. One or more isolates from each cross are then allowed to segregate. Finally, segregants are scored for the presence of the marker to be mapped. If the insertion is found in only a fraction of the segregants, it maps within the duplicated region (Fig. 2). If all segregants contain the insertion, it maps outside the duplication interval.

This procedure can be applied to any dominant marker. In the protocol below, we describe its use for mapping Tn*10* insertions, which can be viewed as a general procedure to map any mutation, provided that a linked Tn*10* insertion is previously obtained.

2   Materials

2.1   Bacterial strains

Collections of *Salmonella enterica* strains are listed in Tables 1 and 2. All belong to serovar Typhimurium and derive from the standard wild-type LT2. Preparation of P22 tails requires use of strain PY13759, which carries plasmid pPB13 (Ap^r). This plasmid constitutively expresses P22 gene *9* [11].

2.2   Bacteriophages

P22 HT 105/1 *int201* is a high-transducing P22 derivative that packages host DNA at a high frequency [4, 18]. Mu*d*P and Mu*d*Q are P22-Mu hybrids that carry a chloramphenicol-resistance marker [10, 11].

2.3   Reagents, culture media, and solutions

L broth (LB) can be obtained from commercial suppliers; a recipe also can be found in [4]. L agar is LB containing 1.5% agar. Bochner-Maloy agar for the selection of Tc^s derivatives of a Tc^r (Tn*10*-carrying) strain can be prepared as described by Maloy et al. [4]. The formulae for E salts and E medium also can be found in [4]. Antibiotics are used at the following concentrations: ampicillin, 90 mg/l (for plasmid-encoded Ap resistance); chloramphenicol, 20 mg/l; and tetracycline, 20 mg/l. Mitomycin C is used at a final concentration of 2.5 µg/ml; the solution is made in deionized water and should be used fresh.

3　　Methods

3.1　Genetic mapping of Tn*10* insertions with locked-in Mu*d*-P22 prophages

Protocol 1　Preparation of Mu*d*-P22 lysates

1. Prepare 2 ml LB cultures of the strains carrying Mu*d*-P22 prophages. Grow until saturation (overnight) at 37 °C, with shaking.
2. Using 2 ml of each saturated culture, inoculate flasks containing 100 ml LB containing 2% glucose and E salts. Incubate 90 min at 37 oC, with shaking.
3. Add 100 µl of a mitomycin C solution, and continue shaking for 3 h (or until lysis is observed).
4. Centrifuge 5–10 min at 5000 rpm to eliminate cell debris.
5. Transfer supernatants to sterile, screw-capped tubes. Add 30 µl of a suspension of P22 tails (see protocol 2 below). Allow head-tail assembly by incubating overnight at room temperature.
6. Add 2 ml chloroform and vortex vigorously. Maintain at room temperature for 1–2 days, and then store at 4 °C.

Protocol 2　Preparation of P22 tails

1. Prepare a culture of strain PY13579 in 3 l LB containing 0.2% glucose, E salts, and ampicillin. Grow until saturation.
2. Harvest the cells by centrifugation (5000 rpm, 15 min).
3. Resuspend the cell pellet in 10 ml Tris-HCl 0.1M pH 5.5, EDTA 10 mM.
4. Add 10 mg lysozyme and incubate at 65 °C for 4 h.
5. Eliminate cell debris by centrifugation (17000 rpm, 70 min).
6. Keep the supernatant at 4 °C for 12–24 h before use. This supernatant may contain up to 10^{14} tails/ml.

An alternative procedure, developed in the laboratory of Stan Maloy, is the use of Mu*d*-P22 lysogens that carry the tail-producing plasmid. Upon induction, tailed phage are directly produced.

Protocol 3   Mapping

1. Prepare an overnight culture of the strain carrying the Tc^r insertion to be mapped.
2. Spread 0.2 ml of the strain onto Bochner-Maloy agar and let it dry.
3. Add drops of undiluted Mud-P22 lysates to obtain a grid pattern in which each donor lysate is identified. Let the plate dry, and incubate at 42 °C.

Interpretation of the results: Whenever a chromosomal fragment is transduced, replacement of the Tc^r marker in the recipient chromosome gives rise to Tc^s transductants and these form a patch on Bochner-Maloy plates. Because the chromosomal location of the donor Mud-P22 prophage is known, the location of the Tc^r marker can be inferred.

3.2   Genetic mapping of Tn*10* insertions by duplication segregation

Protocol 4   Preparation of "P22 broth"

Mix 100 ml nutrient broth, 2 ml 50 x E salts, 1 ml 2% glucose, and 0.1 ml of a phage lysate grown on the wild-type (e. g., LT2). After an optimal infection, such lysates contain up to 10^{11} plaque-forming units per ml. Phage titration by plaque counts is advised.

Protocol 5   Construction of merodiploids carrying the duplication to be mapped

1. Lysate preparation: Mix 4 ml "P22 broth" with 1 ml of a saturated culture of the donor strain. Incubate 8–12 h at 37 °C with gentle shaking.
2. Phage harvest: Centrifuge 15 min at 4500 rpm, save supernatant, and transfer to a screw-capped tube. Add a few drops of chloroform and vortex. Keep overnight at room temperature and then store at 4 °C.
3. Transduction: Mix 0.1 ml aliquots of the phage lysate (diluted as needed) with 0.1 ml of a saturated culture of each recipient, grown in nutrient broth with chloramphenicol. Incubate the mixtures 30 min at 37 °C with shaking to allow phage adsorption. Spread on L agar plates supplemented with tetracycline and chloramphenicol. Incubate at 37 °C for 24–36 h.

Protocol 6   Mapping

1. Grow 3–4 individual Tc^r Cm^r transductants from each cross to full density in L broth to permit segregation.
2. Dilute the cultures and spread aliquots on L agar plates to obtain single colonies.

3. Using sterile toothpicks, patch 100 colonies to the following media: (1) L agar with chloramphenicol and tetracycline to detect non-segregating Cm^r Tc^r colonies; (2) L agar with tetracycline to detect the presence of segregants (which are Cm^s); and (3) L agar without antibiotics to detect Cm^s Tet^s segregants (in which the duplication and the Tn*10* element have co-segregated).

Interpretation of the results: If all the Cm^s segregants contain the transduced marker, it maps outside the duplication. If the marker is found in only a fraction of the segregants, it maps within the duplicated region.

4   Troubleshooting

- Packaging by a given locked-in Mu*d*-P22 hybrid prophage is mostly unidirectional, but residual bidirectional packaging may also occur. In such cases, lysates from Mu*d*P and Mu*d*Q elements inserted at the same chromosomal location will contain overlapping chromosomal stretches, and mapping will be slightly less precise.
- Mapping by duplication segregation may encounter the problem that duplications affecting certain chromosomal regions segregate slowly even in the absence of selection [16]. In such cases, the number of Cm^r segregants found upon non-selective growth can be insufficient to score co-segregation of the transduced marker. To solve this problem, longer segregation periods can be allowed (e. g. 40–60 generations, which can be achieved by two or more serial cultures in nutrient broth). Slow segregation is the only potential complication of the method, which is otherwise unambiguous.

5   Remarks and conclusions

Identification of mutations that cause hitherto unknown phenotypes or affect uncharacterized loci can hardly be addressed directly with molecular biology methods. The problem becomes specially insidious when point mutations are involved (and, in general, whenever a mutation is not tagged with a known marker). In such cases, localization of the mutation on a genetic map is a necessary step previous to molecular analysis. Genetic mapping also can facilitate the construction of isogenic strains, the construction of strains with different combinations of markers, and the localized mutagenesis of single-copy genes. For such purposes, the mapping procedures developed in *Salmonella* are technically simple and fast. Actually, the most time-consuming step to map a point mutation on the *Salmonella* chromosome is the introduction of a neighboring insertion to be used as a landmark. For this purpose, Tn*10* derivatives have the advantage of being both selectable and counterselectable [4, 13, 14]. A potential limitation of Tn*10*, its relatively high specificity of insertion, can be overcome by use of transposase variants with broader target specificity [19]. Once a Tn*10* element has been placed near the mutation, mapping can be readily performed, usually in two steps: (1) the duplication mapping procedure can ascribe the mutation to a chromosome segment; (2) mapping within the segment can be then carried out with locked-in Mu*d*-P22 prophages, to place the marker within 1–4 centisomes. Lysates obtained by induction of locked-in Mu*d*-P22 prophages are also good sources of DNA for cloning and may facilitate physical analysis of the locus under study [20].

Aside from mapping, chromosomal duplications with known endpoints can be used for other operations of genetic analysis. For instance, the introduction of a mutation into one copy of a chromosomal duplication permits complementation analysis: the phenotype of a dominant mutation will still be observed in the merodiploid, while that of recessive mutation will be observed only in haploid segregants. An advantage of this procedure (e. g., over-complementation with a plasmid-borne, wild-type allele) is that it does not alter gene dosage. Duplications also can be useful to ascertain whether a mutation is lethal. For this purpose, a plasmid-borne allele can be recombined into one copy of a duplication; the segregation pattern of the resulting heterozygote will indicate whether haploid segregants carrying the mutation are viable. Lastly, duplications also can be used for the analysis of gene fusions, especially for the study of genes that regulate their own transcription. In such cases, use of a duplication permits one to monitor gene expression in the presence of the normal dosage of the wild-type allele.

Acknowledgments

Our work is supported by grants from the Spanish Ministry of Science (BIO2001–0232-CO2–02) and the European Union (QLK2–1999–0310). We are grateful to Stan Maloy for critical reading of the manuscript.

References

1   Zinder ND, Lederberg J (1952) Genetic exchange in *Salmonella*. *J Bacteriol* 64: 679–699

2   Sanderson KE, MacLachlan PR (1987) F-mediated conjugation, F⁺ strains, and Hfr strains of *Salmonella typhimurium* and *Salmonella abony*. In: FC Neidhardt, JL Ingraham, KB Low et al. (eds): *Escherichia coli and Salmonella typhimurium:*

Cellular and molecular biology. ASM Press, Washington DC, 1138–1144

3 Margolin P (1987) Generalized transduction. In: FC Neidhardt, JL Ingraham, KB Low et al. (eds): *Escherichia coli and Salmonella typhimurium: Cellular and molecular biology.* ASM Press, Washington D C, 1154–1168

4 Maloy SR, Stewart VJ, Taylor RK (1996) *Genetic analysis of pathogenic bacteria.* Cold Spring Harbor Laboratory Press, Cold Spring Harbor, New York

5 Ahmer BMM, Tran M, Heffron F (1999) The virulence plasmid of *Salmonella typhimurium* is self-transmissible. *J Bacteriol* 181: 1364–1368

6 Smith HR, Humphreys GO, Grindley NDF, Anderson ES (1973) Molecular studies of a fi+ plasmid from strains of *Salmonella typhimurium* LT2. *Mol Gen Genet* 126: 143–151

7 Lam S, Roth JR (1983) IS*200*: a *Salmonella*-specific insertion sequence. *Cell* 34: 951–960

8 Chumley FG, Menzel R, Roth JR (1979) Hfr formation directed by Tn*10*. *Genetics* 91: 639–655

9 Chumley FG, Roth JR (1980) Rearrangement of the bacterial chromosome using Tn*10* as a region of homology. *Genetics* 94: 1–14

10 Benson NR, Goldman B S (1992) Rapid mapping in *Salmonella typhimurium* with Mu*d*-P22 prophages. *J Bacteriol* 174: 1673–1681

11 Youderian P, Sugiono P, Brewer KL, Higgins NP, Elliott TE (1988) Packaging specific fragments of the *Salmonella* chromosome with locked-in Mu*d*-P22 prophages. *Genetics* 118: 581–592

12 Casjens S, Hayden M (1988) Analysis *in vivo* of the bacteriophage P22 headful nuclease. *J Mol Biol* 194: 411–422

13 Bochner BR, Huang HC, Schieven GL, Ames BN (1980) Positive selection for loss of tetracycline resistance. *J Bacteriol* 143: 926–933

14 Maloy SR, Nunn WD (1981) Selection for loss of tetracycline resistance by *Escherichia coli. J Bacteriol* 145: 1110–1112; erratum 146: 831

15 Kleckner N, Roth J, Botstein D (1977) Genetic engineering *in vivo* using translocatable drug-resistant elements. New methods in bacterial genetics. *J Mol Biol* 116: 125–159

16 Camacho EM, Casadesus J (2001) Genetic mapping by duplication segregation in *Salmonella enterica. Genetics* 157: 491–502

17 Hughes KT, Roth JR (1985) Directed formation of deletions and duplications using Mu*d*(Ap, *lac*). *Genetics* 109: 263–282

18 Schmieger H (1972) P22 mutants with increased or decreased transduction abilities. *Mol Gen Genet* 119: 75–88

19 Kleckner N, Bender J, Gottesman S (1991) Uses of transposons with emphasis on Tn*10. Meth Enzymol* 204: 139–180

20 Beuzon CR, Casadesus J (1997) Cloning with Mu*d*-P22 hybrid prophages: mapping of IS*200* elements on the chromosome of *Salmonella typhimurium* LT2. *Mol Gen Genet* 256: 586–588

3 Insertion Sequences as Genomic Markers

Dominique Schneider and Michel Blot

Contents

1   Introduction

Mobile genetic elements are widespread in almost all living organisms. This chapter will focus on insertion sequence (IS) elements, which are bacterial mobile genetic elements carrying genetic information devoted to their transposition and its regulation [1]. IS elements, and more generally mobile genetic elements, were first discovered by their ability to generate mutations [2, 3], and several studies suggested their significant contribution to spontaneous mutagenesis in bacteria [4, 5]. Transposition of an IS element can result in gene inactivation, polar effects [6, 7], activation of cryptic genes, and modification of gene expression (for review, see [1]). Besides these "simple" transposition events, IS elements can be involved in global restructuring of genomes, through

Methods and Tools in Biosciences and Medicine
Prokaryotic Genomics, ed. by M. Blot
© 2003 Birkhäuser Verlag Basel/Switzerland

homologous recombination events between homologous copies. Chromosomal rearrangements such as inversions, deletions, and duplications have been described [8, 9].

IS elements therefore generate a wide range of mutations, potentially leading to a modification of gene expression and a restructuring of genomes. These mutations give rise to a strong polymorphism in bacterial populations, which might be of considerable importance for the evolution of bacteria. Moreover, IS elements can be found located onto horizontally transferable DNA [8], e. g., conjugative plasmids or pathogenicity islands. Examples of IS elements forming composite transposons carrying catabolic genes are known [10]. These genes allow metabolism of organic compounds present in the environment and might allow evolutionary changes in bacteria by horizontal transfer. In some cases, horizontal transfer of adhesin loci into *Escherichia coli* involves IS elements [11].

As a result of replicative transposition, the insertion events also may lead to an increased copy number of IS elements, allowing generation of numerous copies in many genomic locations. Analysis of the distribution of IS elements in natural isolates of *E. coli* suggested a rapid change in copy numbers and positions during evolution [12]. The high variability in copy number and location in these natural isolates probably reflects both horizontal transfer of IS elements between strains and transposition within a host. Dynamics of IS elements was also studied between clinical isolates of *Mycobacterium tuberculosis* for epidemiological analyses. Although a remarkably restricted allelic diversity was observed in *M. tuberculosis* [13], probably reflecting a recent evolutionary bottleneck, a high level of genetic diversity was revealed by using an IS element as probe [14].

Extensive intrastrain diversity also can be detected. Restriction fragment length polymorphism (RFLP) analyses of bacterial clones recovered from a 30-year-old stab culture of *E. coli* W3110 revealed high polymorphism [15]. On average, each clone revealed about 12 changes from the putative ancestor, with a mutation rate of about 10^{-5} IS-related DNA rearrangement per bacterial chromosome per hour of storage [16].

IS elements are highly mutagenic and generate a wide range of mutational events that alter the pattern of gene expression and reorganize the genome structure. Therefore, they might constitute good genotypic markers for monitoring of the evolution of bacterial populations. It is usually not possible to precisely follow and investigate the generation and maintenance of diversity during significantly long evolutionary times. Consequently, it is not always experimentally easy to compare rates of phenotypic and genomic changes. Another important challenge is to identify, among the generated polymorphism, some mutations responsible for the adaptive phenotypic changes occurring during evolution. This chapter describes the use of IS elements during experimental evolutions of *E. coli* to measure genetic diversity through time. Mapping of the IS insertions was also used to identify mutations that were fixed early in the populations and therefore represent good candidates to confer a selective advantage.

2   Materials

Equipment
- Bench centrifuge for Eppendorff tubes (up to 13,000 rpm), spectrophotometer, cycler for polymerase chain reaction (PCR).

Chemicals
- Luria broth (LB) medium: 20 g/l tryptone, 10 g/l yeast extract, 10 g/l NaCL.
- Lysozyme (Sigma), agarose (Sigma).

Solutions, reagents, and buffers
- Lysis solution: 25 mM Tris-HCl (pH 7.4), 50 mM glucose, 10 mM EDTA, 2 mg/ml lysozyme.
- 20% sodium dodecyl sulfate.
- 0.3 M sodium acetate-saturated phenol, 3 M sodium acetate.
- Ethanol (100%, 70%).
- TAE buffer: 0.04 M Tris-acetate, 0.001 M EDTA. Prepare a 50× concentrated stock solution.
- *Eco*RV and *Hinc*II restriction enzymes (Life Technologies), T4 DNA ligase (Roche), High-fidelity Expand Taq DNA polymerase (Roche), DIG High Prime DNA Labeling and Detection Kit (Roche), pCRII-Topo Cloning Kit (Invitrogen).

3   Methods

Protocol 1   Preparation of genomic DNA of *E. coli* strains

This protocol is adapted from [17].
1. Cells from an overnight culture in LB medium were collected by centrifugation and resuspended in 0.5 ml of lysis solution.
2. Cells were lysed by addition of 0.05 ml of 20% sodium dodecyl sulfate, and DNA was extracted with an equal volume of phenol saturated with 0.3 M sodium acetate.
3. The aqueous phase was retained and DNA was precipitated by addition of one tenth of volume of 3 M sodium acetate and 2 volumes of absolute ethanol.
4. The DNA precipitate was washed with 0.1 ml of 70% ethanol.
5. After removal of the supernatant, the DNA was air-dried 10 min and resuspended in 0.1 ml sterile water.
6. Quantitation of DNA was done by spectrophotometry at an optical density (OD) of 260 nm (1 $OD_{260\,nm}$ corresponds to 50 ng/μl of DNA).

Protocol 2 DNA digestion, transfer onto nylon membranes and hybridization experiments

1. Genomic DNA (4 µg) was digested with either *Eco*RV or *Hinc*II for 4 h at 37 °C, according to the manufacturer's recommendations.
2. The obtained restriction fragments were separated by electrophoresis on a 0.8% agarose gel in Tris-Acetate EDTA (TAE) buffer.
3. DNA was subsequently transferred on a nylon membrane (Roche) [18].
4. The probes consisted of internal fragments of IS elements [15]. They were labeled with the DIG High Prime DNA Labeling and Detection Kit from Roche, as recommended by the manufacturer.
5. The nonradioactive system uses digoxigenin (DIG), a steroid hapten, to label DNA for hybridization and subsequent luminescence detection. Digoxigenin is coupled to dUTP, and DIG-labeled DNA probes are generated according to the random primed labeling technique.
6. Hybridization experiments were done at high stringency (68 °C), according to the manufacturer's recommendations.
7. The membranes with the hybridized probes are immunodetected with anti-DIG, Fab fragments are conjugated to alkaline phosphatase and are then visualized with the chemiluminescence substrate CSPD (disodium 3-(4-methoxyspiro(1,2-dioxetane-3,2'-(5'-chloro)tricyclo[3,3.1.1$^{3.7}$]decan}-4-yl)-phenyl phosphate). Enzymatic dephosphorylation of CSPD by alkaline phosphatase leads to a light emission that is recorded on X-ray films. Membranes can be stripped and subsequently used with different probes.

Protocol 3 Phylogenetic methods

1. Every clone was scored for the presence or absence of each fragment that hybridized with a particular IS probe.
2. Ambiguous fragments of similar size usually were resolved by running the relevant clones in parallel and, otherwise, were scored conservatively.
3. Phylogenies were constructed by using a parsimony method in which the roots of the trees were forced to be the actual common ancestor [19]. The purpose of the phylogenies is to illustrate the divergence of the clones from their ancestor and from one another.

Protocol 4 Characterization of sequences adjacent to IS elements by inverse PCR

1. We refer to the left and right sides of an IS element according to the direction of transcription of its transposase gene. Clones from the two evolution populations Ara-1 and Ara+1 (see "Results and discussion") were analyzed by RFLP experiments using *Eco*RV- or *Hinc*II-digested genomic DNA. In the different examples described, significant differences in RFLP experiments were observed for IS*1*, IS*150*, and IS*186*. Therefore, only mapping experiments by inverse PCR will be described here for these

three elements. The same strategy can, however, be applied for any IS element, the critical step being the choice of the primers for the inverse PCR. To map IS*1* and IS*186* elements, genomic DNA was digested with *Eco*RV, whereas *Hinc*II was used for IS*150* mapping. The chosen enzyme should not cut within the IS element.

2. Genomic DNA of a clone was digested with *Eco*RV or *Hinc*II, and fragments were separated onto a 0.8% agarose gel with *Pst*I- and *Hind*III-digested lambda DNA as size markers.

3. Gel fractions containing IS fragments were cut and DNA was purified using the Gel Extraction Kit of Qiagen.

4. These fragments were self-ligated with T4 DNA ligase (Roche) at 5–10 µg/ ml, and the ligated mixtures were used as templates in PCR experiments.

5. Primers used for inverse PCR to amplify sequences adjacent to IS*1* were G3, 5'-GTCATCGGGCATTATCTGAAC-3' and G4, 5'-AGAAGCCACTGGAGCACC-3'. The IS*150* primers were G5, 5'-GATCCTGTAACCATCATCAG-3' and G6, 5'-CTGAAGGATGCTGTTACGG-3'. The IS*186* primers were G7, 5'-CGGCAT-TACGTGCCGAAG-3' and G8, 5'-GGTGGCCATTCGTGGGAC-3'. All primers are near the corresponding IS extremities and directed outward.

6. Sequences of adjacent DNA were done by using the same primers as in the PCR experiments.

7. Sequences were compared to databases by using the BLAST program [20].

8. All adjacent sequences were used as DIG-labeled probes with reference membranes to confirm that predicted IS-containing fragments hybridized. Reference membranes carry *Eco*RV- and *Hinc*II-digested genomic DNA from the ancestor and diverse clones, including those from which the adjacent sequences were obtained.

9. The membranes were probed with the IS elements, stripped, and reprobed with adjacent sequences to show that the correct sequences were cloned.

4   Results and discussion

Evolutionary processes were extensively studied in bacterial populations during the last two decades because of their short generation times and their relatively small genomes. Genetic diversity was monitored in experimental evolution populations of *E. coli* by using IS elements as genotypic markers. Diversity was generated and maintained in these bacterial populations through competition among clones and negative frequency-dependent selection. IS elements also allowed the identification of potentially beneficial mutations. This section will first describe the genomic evolution in experimental *E. coli* populations revealed by IS elements. In the second part, we will describe IS-mediated mutations in these populations. In a third part, we will review some of the experimental evolutions in which IS elements led to beneficial mutations.

4.1   10,000 generations of experimental evolution in *E. coli*

Twelve replicate populations of *E. coli* B, initiated from a single ancestral strain, were serially propagated at 37 °C in Davis minimal medium with a limiting amount of glucose (25 µg/ml) [21]. A daily 100-fold dilution was realized in the same fresh medium. The dilution and regrowth allowed 6.64 generations per day. These populations have now been propagated for about 30,000 generations [22]. The ancestor clone is strictly asexual. Therefore, genetic diversity can be generated only through chromosomal mutations.

We measured genetic diversity and genomic evolution through time by RFLP using the seven known IS elements as probes [23]. Seven to 20 clones were randomly chosen at several time points in two populations, called Ara-1 and Ara+1. Each clone's genomic DNA was hybridized successively with the probes for IS elements. Thus, RFLP is a genetic fingerprint based on the presence or absence of each fragment that hybridizes. Genetic distances among clones were computed and clonal phylogenies were constructed (Fig. 1). They showed successive groups of clones replacing each other and specific individuals harboring "pivotal" mutations that are present in all descendents. A mutation may have reached high frequency either because it conferred a selective advantage or because it hitchhiked with another beneficial mutation [23].

Over time, the two populations diverged increasingly from their common ancestor, both phenotypically and genetically. All these results, obtained by RFLP-IS, highlighted the importance of selection and competition on the genomic evolution of asexual organisms. Selective sweeps of beneficial mutations during the first and rapid adaptation period would purge much of the existing variation, while genetic diversity within populations reached its highest levels only after the decrease in adaptive evolution.

The effect of selection and competition was reflected by the phylogenetic trees derived from the IS fingerprints [23]. These trees contained a main trunk, along which were observed several pivotal genotypes that are ancestral to all individuals subsequently sampled. Every IS-associated mutation defining a pivotal genotype either hitchhiked with a beneficial mutation or was itself beneficial.

In these experimental populations, IS elements were good genotypic markers. Other repeated sequences, such as microsatellites, were also used successfully to study population genetics [24, 25]. However, in contrast to microsatellites, IS fingerprints have directly identified potential beneficial mutations. Mapping these IS insertions might therefore give precious information about the targets of natural selection.

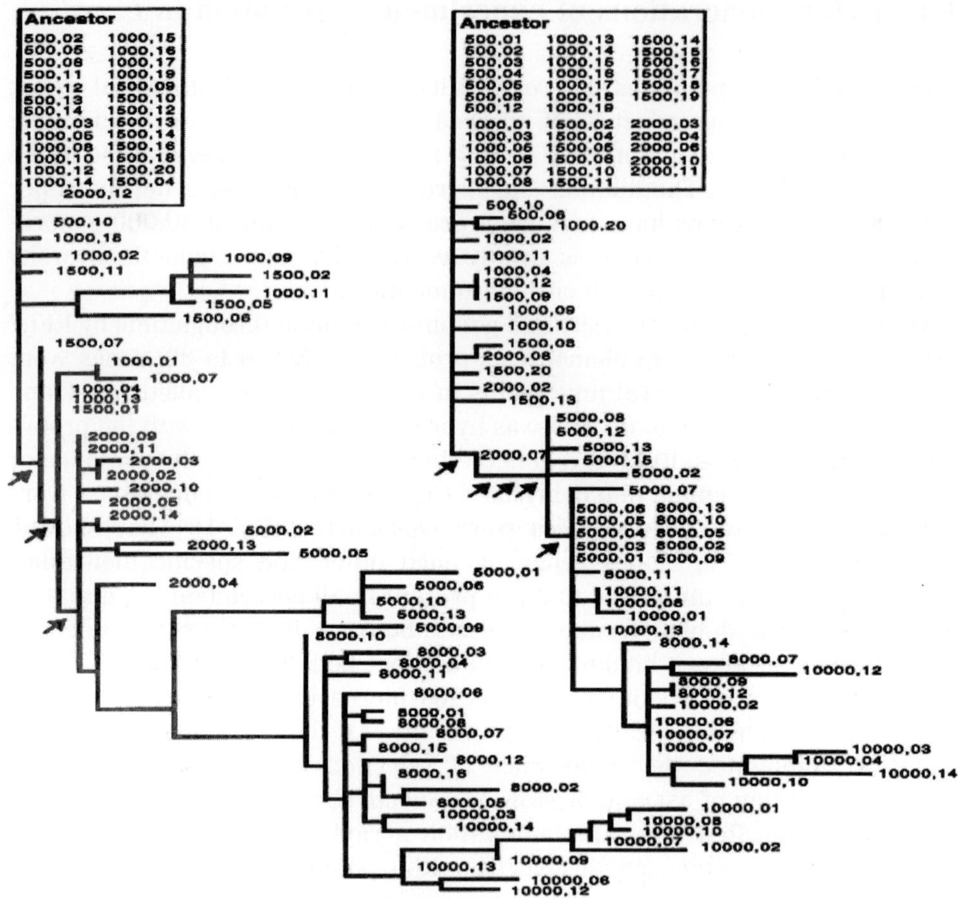

Figure 1   Phylogenies for clones from two evolving populations of *E. coli*, rooted by using the ancestral genotype.
The figure was reproduced from [23]. Phylogenies were inferred by parsimony from IS-RFLP data. Notation indicates the generation at which each clone was sampled, followed by an arbitrary number to distinguish clones from the same sample. Clones in the box with the ancestor were identical to the ancestor on the basis of their IS fingerprints. Arrows mark some of the pivotal mutations that were shared by all clones in every later sample. (A) Population Ara+1. (B) Population Ara-1.

4.2   Mapping the pivotal IS-linked mutations

From the IS fingerprints, we characterized 10 pivotal mutations [26, 27]. Three arose and were fixed in population Ara-1, six in Ara+1, and one was found to be fixed in the 12 populations. All 10 mutations were characterized by determining the insertion sites of the relevant IS elements. This was done by inverse PCR using primers directed outwards from the relevant IS element, as described in the Methods section. All pivotal mutations were IS-mediated events: six simple insertions and four complex rearrangements. The different mutations are described in Table 1. In each of the two populations, a large inversion involving one-third of the chromosome was found, in one case by homologous recombination between two IS*1* elements and in the other case, between two IS*150* elements. An additional deletion, also involving transposition of IS*1*, was characterized in Ara-1. The simple insertions found in Ara-1 and Ara+1 corresponded to transposition events of IS*150*. The target genes for IS-mediated mutations encode functions that seem to be relevant to the environment of this experimental evolution, i. e. minimal glucose medium (Tab. 1).

Table 1 Description of the pivotal mutations detected with IS elements during 10,000 generations of experimental evolution of *E. coli*. Data are from [27].

Mutational event	Population	Gene function	Mutation first detected at (generation)	Mutation fixed by (generation)
rbs deletion	All 12	Ribose utilization	500	2000
pykF::IS*150*	Ara-1	Pyruvate kinase I	2000	5000
Inversion between two IS*1*s (14.1' and 46.9')	Ara-1		5000	5000
Deletion between two IS*1*s	Ara-1		5000	5000
nadR::IS*150*	Ara+1	Repressor of NAD biosynthetic genes	1000	2000
hokB-sokB::IS*150*	Ara+1	Homologous to plasmid stabilization systems	1000	2000
IS*150* insertion in regulatory region of *pbpA*	Ara+1	Penicillin-binding protein 2	2000	2000
Inversion between two IS*150*s (62' and 99.7')	Ara+1		2000	2000
yfcU::IS*150*	Ara+1	unknown	5000	8000
Insertion of IS*150* in intergenic region of *glcB* and *yghK*	Ara+1	Malate synthase G and unknown	5000	8000

The last pivotal mutation was detected in all 12 populations and was fixed very early (500 to 2000 generations). It corresponded to the deletion of the *rbs* operon, which is responsible for ribose utilization [27]. These deletions occurred during transposition of IS*150*. A strain with a deletion of the *rbs* operon was constructed, and it showed a significant competitive advantage over the ancestor clone [27]. The same reconstruction experiments are in progress for the other pivotal mutations.

In conclusion, IS elements allowed us to follow the genetic diversity (generation and maintenance) during 10,000 generations of evolution in *E. coli*. Among the numerous generated mutations, one was shown to be beneficial and four are currently under investigation (unpublished data).

4.3   IS mutations in other experimental evolution populations

In the mid-1980s, a series of selection experiments was started in which *E. coli* populations were propagated for hundreds of generations in glucose-limited chemostat cultures [28]. Despite a single homogeneous environment, stable polymorphisms were generated and maintained in these populations. Three genetically distinct clones coexisted: one of them was better at sequestering glucose and secreted glycerol and acetate. The second one was able to utilize the secreted acetate, and the third was able to use glycerol [29]. The polymorphism was explained by cross-feeding coupled with a difference in glucose-uptake kinetics. Moreover, competition experiments revealed that no single genotype was able to drive another to extinction, leading in all cases to stable equilibria. Measures of acetyl CoA synthetase activities revealed that the acetate cross-feeding phenotype was associated with semi-constitutive over-expression of the *acs* gene, encoding acetyl CoA synthetase [30]. The mutations were localized in the regulatory region of the gene: in two populations out of six, the mutations corresponded to insertions of IS*3* and IS*30*.

Another experimental evolution study used lines of *E. coli* adapted for 2000 generations at different incubation temperatures [31]. Six of these lines, adapted at high temperature (41.5 °C), were analyzed by DNA high-density arrays [32]. Comparison of the intensity of hybridization signals was computed to detect duplication and deletion events. Whereas three lines contained no such events, three duplication and deletion events were detected in one line and one duplication event was found in two lines. An interesting finding was the recovery of a duplication of the same chromosomal locus in three independent lines. The time intervals during which these duplications were fixed in the relevant lines corresponded to a significant increase in fitness, suggesting their adaptive involvement [32]. IS elements, as well as other repeats, were associated with these rearrangements.

Although alternative approaches were used in these later examples, direct mapping of IS elements would also have identified the targets for natural selection. Therefore, we suggest that IS elements as genotypic markers might allow the analysis of population genetics mechanisms giving rise to adaptation.

5   Remarks and conclusions

The use of IS elements as genomic markers addressed several important questions for evolutionary biologists. Measuring genomic diversity in experimental populations with IS elements highlighted the importance of competition and selection in the adaptive strategy of bacteria. The 12 populations created by Richard Lenski clearly highlighted the efficiency of IS elements to reveal diversification both within and across populations. Potential beneficial mutations mediated by IS elements were identified and are currently under study. These analyses will help to clarify whether the different populations found different or identical genetic solutions to achieve the same fitness increase. The role of chance can then be addressed.

IS elements were shown to generate high diversity in pathogenic bacteria and were used to classify them into genetically distinct groups [13]. It was proposed that IS transposition might play an important role in *M. tuberculosis* evolution by altering gene expression [13]. The RFLP-IS data allowed the construction of phylogenetic trees, whose analysis suggests that the genome of clinical isolates is evolving. Therefore, RFLP-IS data might not only give insights into the evolutionary history of pathogenic bacteria but also provide information about pathogenicity. For instance, mapping IS-mediated mutations might enable the identification of mutations conferring an advantage to strains of *M. tuberculosis*.

Acknowledgments

The work described here in two of the 12 *E. coli* populations was done in collaboration with the group of Richard Lenski. We thank Richard Lenski and all his lab members for this stimulating collaboration.

References

1   Mahillon J, Chandler M (1998) Insertion sequences. *Microbiol Mol Biol Rev* 62: 725–774

2   McClintock B (1965) The control of gene expression in maize. *Brookhaven Symp Biol* 18: 162–184

3   Starlinger P, Saedler H (1976) IS-elements in microorganisms. *Curr Top Microbiol Immunol* 75: 111–152

4   Lieb M (1981) A fine structure map of spontaneous and induced mutations in the lambda repressor gene, including insertions of IS elements. *Mol Gen Genet* 184: 364–371

5   Hall BG (1999) Spectra of spontaneous growth-dependent and adaptive mutations at *ebgR*. *J Bacteriol* 181: 1149–1155

6   Jordan E, Saedler H, Starlinger P (1968) O° and strong polar mutations in the *gal* operon are insertions. *Mol Gen Genet* 102: 353–363

7   Saedler H, Reif HJ, Hu S, Davidson N (1974) IS*2*, a genetic element for turn-off and turn-on of gene activity in *Escherichia coli. Mol Gen Genet* 132: 265–289

8   Deonier RC (1996) Native insertion sequence elements: locations, distributions, and sequence relationships. In: FC Neidhardt (eds): *Escherichia coli and Salmonella: Cellular and molecular biology*. ASM Press, Washington, DC, 2000–2011

9   Louarn J-M, Bouche JP, Legendre F et al. (1985) Characterization and properties of very large inversions of the *E. coli*

chromosome along the origin-to-terminus axes. *Mol Gen Genet* 201: 467–476

10 Fong KPY, Goh CBH, Tan H-M (2000) The genes for benzene catabolism in *Pseudomonas putida* ML2 are flanked by two copies of the insertion element IS*1489*, forming a class-I-type catabolic transposon, Tn*5542*. *Plasmid* 43: 103–110

11 Garcia MI, Labigne A, Le Bouguenec C (1994) Nucleotide sequence of the afimbrial-adhesin-encoding *afa-3* gene cluster and its translocation via flanking IS*1* insertion sequences. *J Bacteriol* 176: 7601–7613

12 Sawyer SA, Dykhuizen DE, DuBose RF et al. (1987) Distribution and abundance of insertion sequences among natural isolates of *Escherichia coli*. *Genetics* 115: 51–63

13 Sreevatsan S, Pan X, Stockbauer KE, et al. (1997) Restricted structural gene polymorphism in the *Mycobacterium tuberculosis* complex indicates evolutionarily recent global dissemination. *Proc Natl Acad Sci USA* 94: 9869–9874

14 Warren RM, Sampson SL, Richardson M, et al. (2000) Mapping of IS*6110* flanking regions in clinical isolates of *Mycobacterium tuberculosis* demonstrates genome plasticity. *Mol Microbiol* 37: 1405–1416

15 Naas T, Blot M, Fitch WM, Arber W (1994) Insertion sequence-related genetic variation in resting *Escherichia coli* K-12. *Genetics* 136: 721–730

16 Naas T, Blot M, Fitch WM, Arber W (1995) Dynamics of IS-related genetic rearrangements in resting *Escherichia coli* K-12. *Mol Biol Evol* 12: 198–207

17 Nichols BP, Shafiq O, Meiners V (1998) Sequence analysis of Tn*10* insertion sites in a collection of *Escherichia coli* strains used for genetic mapping and strain construction. *J Bacteriol* 180: 6408–6411

18 Southern EM (1975) Detection of specific sequences among DNA fragments separated by gel electrophoresis. *J Biol Chem* 98: 503–517

19 Swofford DL (1993) *PAUP: Phylogenetic analysis using parsimony.* Illinois Natural History Survey, Champaign, IL, Version 3.1.1

20 Altschul S, Stephen F, Madden TL, et al. (1997) Gapped BLAST and PSI-BLAST: a new generation of protein database search programs. *Nucleic Acids Res* 25: 3389–3402

21 Lenski RE, Rose MR, Simpson SC, Tadler SC (1991) Long-term experimental evolution in *Escherichia coli*. I. Adaptation and divergence during 2,000 generations. *Am Nat* 138: 1315–1341

22 Lenski RE, Mongold JA, Sniegowski PD, et al. (1998) Evolution of competitive fitness in experimental populations of *Escherichia coli*: what makes one genotype a better competitor than another? *Antonie Leeuwenhoek* 73: 35–47

23 Papadopoulos D, Schneider D, Meier-Eiss J et al. (1999) Genomic evolution during a 10,000-generation experiment with bacteria. *Proc Natl Acad Sci USA* 96: 3807–3812

24 Imhof M, Schlötterer C (2001) Fitness effects of advantageous mutations in evolving *Escherichia coli* populations. *Proc Natl Acad Sci USA* 98: 1113–1117

25 Metzgar D, Thomas E, Davis C, Field D, Wills C (2001) The microsatellites of *Escherichia coli:* rapidly evolving repetitive DNAs in a non-pathogenic prokaryote. *Mol Microbiol* 39: 183–190

26 Schneider D, Duperchy E, Coursange E et al. (2000) Long-term experimental evolution in *Escherichia coli*. IX. Characterization of insertion sequence-mediated mutations and rearrangements. *Genetics* 156: 477–488

27 Cooper VS, Schneider D, Blot M, Lenski RE (2001) Mechanisms causing rapid and parallel losses of ribose catabolism in evolving populations of *Escherichia coli* B. *J Bacteriol* 183: 2834–2841

28 Helling RB, Vargas CN, Adams J (1987) Evolution of *Escherichia coli* during growth in a constant environment. *Genetics* 116: 349–358

29 Rosenzweig RF, Sharp RR, Treves DS, Adams J (1994) Microbial evolution in a simple unstructured environment: genetic differentiation in *Escherichia coli*. *Genetics* 137: 903–917

30 Treves DS, Manning S., Adams J (1998) Repeated evolution of an acetate-cross-feeding polymorphism in long-term populations of *Escherichia coli*. *Mol Biol Evol* 15: 789–797

31 Bennett AF, Lenski RE, Mittler JE (1992) Evolutionary adaptation to temperature. I. Fitness responses of *Escherichia coli* to changes in its thermal environment. *Evolution* 46: 16–30

32 Riehle MM, Bennett AF, Long AD (2001) Genetic architecture of thermal adaptation in *Escherichia coli*. *Proc Natl Acad Sci USA* 98: 525–530

The Use of Noncoding Microsatellite Length Analysis for Bacterial Strain Typing

David Metzgar

Contents

1   Introduction

Microsatellites have found great utility as a primary source of population genetics data for higher eukaryotes [1]. These sequences, defined as tandem repeats with motif lengths of one to six base pairs, are hypermutable (and hypervariable), with mutation rates several orders of magnitude higher than base substitution rates. This hypermutability is thought to derive from the process of replication slippage, in which repeats mispair during transient nonprocessive periods of DNA replication, resulting in a loss or gain of one (or, less commonly, more than one) whole repeat unit [2, 3]. This unique mutational process creates a unique type of variability, in which microsatellite sequences vary between lineages in terms of repeat number. It is this variation that has been used widely in both paternity analysis and population differentiation in studies of higher eukaryotes and yeast.

Methods and Tools in Biosciences and Medicine
Prokaryotic Genomics, ed. by M. Blot
© 2003 Birkhäuser Verlag Basel/Switzerland

Microsatellites are extremely common in eukaryotic genomes, yet they are virtually absent from prokaryotic genomes, despite the fact that the process (replication slippage) thought to be responsible for their ubiquity in eukaryotes also appears to occur in prokaryotes [2, 4]. This distributional difference appears to result from mutation-rate biases that favor repeat expansion (insertion) in eukaryotes and repeat contraction (deletion) in prokaryotes [5]. These biases may have resulted from selection that favors minimal genome-size in prokaryotes.

Prokaryotic genomes do, however, contain some microsatellite sequences. The majority of these are functional repeats associated with genes (contingency loci) that switch at a high rate between functional and nonfunctional states through length mutations in the repetitive sequences that disrupt promoter or coding regions. These mutations generate adaptive variability in bacteria that occupy predictably variable niche spaces and, as such, are thought to be subject to strong selection for both specific character states and for length in general [6]. Contingency microsatellites have been found primarily in pathogens. The most obvious examples are responsible for rapid antigenic switching that allows pathogens such as *Haemophilus influenzae* and *Neisseria gonorrhoeae* to evade host immune responses [7, 8]. Such repeats may be identifiable by their position within promoter/operator regions or within coding sequences, positions which suggest that length changes would result in changes in gene expression or function (see [9]). Because these microsatellites control adaptive variation, they can change very rapidly in strains under strong selection and are expected to change in a manner directly correlated with their immediate environment. As such, these repeats need to be avoided in attempts to use repeat variation as a strain-identification tool.

There are microsatellites in prokaryotes that do not appear to be functional and appear to evolve in a neutral fashion. In our own studies of microsatellite variation in *Escherichia coli*, we identified more than 100 mononucleotide runs in apparently nonfunctional regions. [10] (See Tab. 1). Despite their short length, the seven of these that we tested showed a high level of interstrain variability, suggesting a potential source of readily analyzable strain markers. Many loci did not amplify in all strains, probably as a result of primer site divergence in the noncoding regions bearing the analyzed microsatellites. The use of these microsatellites as markers is very simple and of low cost and is described in detail below. There are several important considerations that need to be made regarding such use, and these are explained in depth in the Troubleshooting section.

Table 1 Characteristics of analyzed *Escherichia coli* repeats. (Table modified with permission from [10]).

Locus	Apparent functional status	Repeat sequence	Allele length relative to K-12 genome and (frequency) A=Atypical (did not amplify)
TNT1	Noncoding	T(9)	−2(1) −1(4) 0(6) +1(1) A(10)
ANTW	Noncoding	A(9)...A(7)...T(7)	−1(6) 0(11) A(5)
OMPF	Noncoding	T(8)	−1(11) 0(10) A(1)
AAS1	Noncoding	A(8)...A(8)	0(7) +1(15)
NCC1	Noncoding	C(9)	−1(4) 0(8) +1(5) +2(1) A(4)
TNT2	Noncoding	T(9)	−1(4) 0(18)
INTG	Noncoding	G(10)	−1(2) 0(3) A(17)

2 Methods

Protocol 1 Microsatellite identification

1. Microsatellites can be identified by using a computer to scan Genbank sequences for long mononucleotide runs (and other types of repeats, though we found no others that matched our requirements for markers in *E. coli*). Nucleotide BLAST searches (http://www.ncbi.nlm.nih.gov/BLAST/) may be used in the absence of more specifically designed programs, though this requires turning off simple sequence filters that will otherwise ignore repetitive microsatellite sequences.
2. Runs can be analyzed individually by reference to Genbank annotations to identify those that are in noncoding (and probably nonfunctional) sequence contexts.
3. Primers are designed to amplify the regions containing the repeats.
4. We attempted to standardize the length and GC content so that common PCR cycling protocols could be designed to allow multiplexing.
5. Primers can be designed to amplify different loci as fragments of different lengths, allowing multiple loci to be analyzed together in single wells. With small differences in length, it is best if the PCR product is as short as possible.

Protocol 2   DNA sample preparation using Chelex (Sigma TM)

1. A single colony of the bacterial strain is picked with a toothpick (pick a small portion if the colony is big) and resuspended in 300 µl of water containing 10% wt/vol 100-mesh Chelex (Sigma-Aldrich) in a 0.6 ml Eppendorf tube. Alternatively, 2–5 µl of liquid culture may be used.
2. This sample is then vortexed for 10 s, boiled for 10 min, vortexed for 10 s again, then centrifuged at 14,000 rpm in a microfuge to pellet any debris. Such samples may be stored for at least a year at –20°C, but they should be thawed, vortexed, and centrifuged before reuse.
3. The remaining Chelex beads at the bottom of the sample must be avoided when removing an aliquot for use in PCR, as the beads will inhibit the reaction. Any other standard DNA prep will, of course, also work, but this is the best method we have used. We have found this method to be extremely easy and cheap, and it works very well for *any* PCR application.

Protocol 3   PCR of microsatellites

It is notable that this protocol has been used in only one study [10] and may not work well in other cases. Protocols similar to those described in this book for general microsatellite analysis and imperfect microsatellite analysis also may be attempted.

Reaction mix:
- 2 mM each dNTP
- 1 µl Pfu buffer (Stratagene)
- 1 unit cloned Pfu DNA polymerase (Stratagene)
- 0.5 µM forward primer
- 0.5 µM reverse primer (multiple primer pairs can often be multiplexed if they give products of
- differing length, though this requires further optimization)
- 0.1 µl fluorescently labeled dCTP (Perkin-Elmer). We used R6G and R110. (the use of different dyes to label different reactions allows multiplexing of
- independently amplified products in single wells of length-analysis gels)
- 1 µl DNA prepared in Chelex (see above)

Amplification:
Standard conditions appropriate to the specific primers, with annealing temperature optimized for each primer pair, should be used.

Protocol 4   Length analysis

Our analysis was performed by using ABI Genescan technology (following manufacturer's protocols from Applied Biosystems) to score the lengths of dye-labeled PCR products. The lengths reported using Genescan technology are approximate and vary in different gels. Therefore, all length analyses must be performed twice on two separate gels, with products from all strains in each comparison being run together on the same gel. Correlations between lengths on independent runs can be used to identify length differences. We accomplished this by creating graphs in which the allele lengths of all samples from one run were represented on one axis, while their lengths on a second run were represented on the second axis. Points representing individuals with alleles of a particular length were then found to cluster as a group along a diagonal with a slope of one through the middle of the graph. Groups with different allele lengths cluster separately.

Length analysis also can be accomplished by sequencing PCR products using standard protocols (for sequences containing A/T runs, when appropriate), and this is likely to give less ambiguous results.

3   Applications

Microsatellite typing in bacteria will never replace the use of markers that provide a greater range of less revertible allelic states for use in phylogenetic reconstruction (such as base polymorphisms and isozymes; see, e. g., [11, 12]) because of relatively high rates of homoplasy (convergence) in microsatellite sequences (see Tab. 2). However, this method can provide a rapid, simple, and inexpensive method to discriminate between multiple strains of a particular bacterial species that coexist in physically limited environments.

Table 2   Phylogenetic information indices of polymorphic noncoding repeats (in the context of the accepted ECOR phylogeny as reported in [11, 12]. Indices calculated in MacClade 3.05 (Maddison and Maddison, 1992). (Table modified with permission from [10]). The most important thing to take from this table is that two strains are more likely to share an allele by chance convergence (homoplasy) (66%) than by ancestry (consistency) (34%).

Locus	AAS1	NCC1	TNT1	ANTW	TNT2	INTG	OMPF	Total
# of steps	4	6	4	5	4	2	7	32
Min. steps	1	3	3	1	1	1	1	11
Max. steps	7	11	7	6	4	2	10	47
Consistency index	0.25	0.5	0.75	0.2	0.25	0.5	0.14	0.34
Retention index	0.5	0.63	0.75	0.2	0	0	0.33	0.42
Rescaled CI	0.13	0.31	0.56	0.04	0	0	0.05	0.14
Homoplasy index	0.75	0.5	0.25	0.8	0.75	0.5	0.8	0.66

This method could be used, for example, to follow the movement of a specific strain of bacteria through a contiguous population of hosts (e. g., a bacterium causing nososcomial infections in a particular hospital), but it would be inappropriate for identifying phylogenetic relationships between bacterial samples from sites all over the world (see [10], in which we show that very divergent *E. coli* strains from the ECOR collection of reference may share identical microsatellite alleles at all seven loci studied, apparently through convergence).

Given that microsatellites generally mutate faster than other types of sequence, these markers may provide sources of variation in cases where other, more widely accepted types of markers are monomorphic, allowing the highest possible degree of strain discrimination.

4   Troubleshooting

It cannot be stressed enough that only nonfunctional variation should be used for purposes of identification. Shared functional repeat lengths may represent shared environmental conditions, rather than shared ancestry, and as such should be avoided in typing efforts. To this end, the genomic context of the repeats must be thoroughly investigated. Also, a brief survey of the population length distribution of any newly identified repeat should be undertaken to be sure that 1) the repeat is variable; 2) the length distribution is approximately normal, as one would expect given neutral evolution from a common ancestor; and 3) the lengths do not appear to be distributed in a manner suggesting evolution in parallel with environmental circumstances.

Microsatellites mutate very rapidly and can easily converge on the same length in different lineages (homoplasy). This is especially true of bacterial repeats because they are generally short and therefore have fewer potential lengths among which to vary. This makes these markers inappropriate for the deduction (or induction) of relatively distant phylogenetic relationships. Rather, they should be used only to discriminate similar strains from one another (identity analysis), and they should always be used in numbers great enough to allow a statistical analysis of correlation between multiple variable markers.

While functional microsatellites have been studied in detail in many prokaryotes, the use of nonfunctional repeats as bacterial strain markers is a very new concept. The density of microsatellites, as well as their rate of mutation and length variability, may differ greatly among bacterial species. Until this method has been used in more studies, it should be considered highly experimental, and considerable optimization and careful, skeptical analysis will be necessary.

Acknowledgments

David Metzgar is an employee of The Scripps Research Institute, La Jolla, CA, USA.

References

1   Schlotterer C, Pemberton J (1994) The use of microsatellites for genetic analysis of natural populations. In: (B Schierwater, B Streit, GP Wagner, R DeSalle, eds.): *Molecular ecology and evolution: Approaches and applications.* Basel, Switzerland: Birkhauser Verlag, 203–214

2   Levinson G, Gutman GA (1987) Slipped-strand mispairing: a major mechanism for DNA sequence evolution. *Mol Biol Evol* 4: 203–221

3   Schlotterer C, Tautz D (1992) Slippage synthesis of simple sequence DNA. *Nucleic Acids Res* 20: 211–215

4   Morel P, Reverdy C, Michel B, Ehrlich SD, Cassuto E (1998) The role of SOS and flap processing in microsatellite instability in *Escherichia coli. Proc Natl Acad Sci USA* 95: 10003–10008

5   Metzgar D, Liu L, Hansen C, Dybvig K, Wills C (2002) Domain-level differences in microsatellite distribution and content result from different relative rates of insertion and deletion mutations. *Genome Res* 12: 408–413

6   Metzgar D, Wills C (2000) Evidence for the adaptive evolution of mutation rates. *Cell* 101: 581–584

7   Moxon ER, Rainey PB, Nowak MA, Lenski RE (1994) Adaptive evolution of highly mutable loci in pathogenic bacteria. *Curr Biol* 4: 24–33

8   Stern A, Brown M, Nickel P, Meyer TF (1986) Opacity genes in Neisseria gonorrhoeae: control of phase and antigenic variation. *Cell* 47: 61–67

9   Field D, Hood D, Moxon R (1999) Contribution of genomics to bacterial pathogenesis. *Curr Opin Genet Dev* 9(6): 700–703

10 Metzgar D, Thomas E, Davis C, Field D, Wills C (2000) The microsatellites of *Escherichia coli*: An analysis of the variability and distribution of short repetitive DNAs in a nonpathogenic prokaryote. *Mol Microbiol* 39: 183–190

11 Herzer PJ, Inouye S, Inouye M, Whitman TS (1990) Phylogenetic distribution of branched RNA-linked multicopy single-stranded DNA among natural isolates of *Escherichia coli. J Bacteriol* 172: 6175–6181

12 Lecointre G, Rachdi L, Darlu P, Denamur E (1998) *Escherichia coli* molecular phylogeny using the incongruence length difference test. *Mol Biol Evol* 15: 1685–16

5 How to Amplify Easily, on the Bacterial Chromosome, a Desired DNA Sequence

Richard D'Ari and Daniel Vinella

Contents

1 Introduction

Gene amplification is a powerful technique with a number of applications. Multicopy suppression, whereby the phenotype caused by loss of one gene product is corrected by increasing another, can provide information on possible protein-protein interactions in the cell [1]. Increased concentrations of particular enzymes can reveal weak activities and possible "underground" reactions with alternative substrates [2]. Amplification of a specific DNA binding site, coupled with expression of the corresponding DNA-binding protein carrying a fluorescence label, has permitted subcellular labelling of specific chromosomal regions [3, 4].

The most widely employed method of gene amplification is by cloning the gene on a multicopy plasmid, although this technique cannot provide chromosomal tags *in cis*, as required for labelling specific chromosomal regions. For other purposes as well, it is sometimes advantageous to have the extra gene copies at their normal chromosomal location. This avoids possible problems

Methods and Tools in Biosciences and Medicine
Prokaryotic Genomics, ed. by M. Blot
© 2003 Birkhäuser Verlag Basel/Switzerland

caused by variation in plasmid copy number in different cells or under different growth conditions.

A number of techniques are available for constructing precise chromosomal deletions or deletion-substitutions [5]. We present here a method for creating precise tandem chromosomal duplications of essentially unlimited length. We also show how to transfer such duplications easily from one strain to another. Finally, we suggest an additional selection by which duplications can give rise to further gene amplification.

These techniques stem from a selection we developed for mutants of *Escherichia coli* that overproduce the cell division proteins FtsQ, FtsA, and FtsZ, thereby becoming resistant to the β-lactam antibiotic mecillinam [6–8]. We originally hoped to identify regulators of the *ftsQAZ* operon. We found instead seven independent mutants in which this operon had been amplified [8]. Our study of the molecular structure and genetic properties of these mutants led to the method of directed specific gene amplification proposed here. It should be stated at the outset, however, that the method is based on our *a posteriori* analysis of these mutants; we have not used it directly to amplify a specific chromosomal fragment chosen *a priori*. Information on troubleshooting is mentioned in the text where it fits the methodological context.

2   Materials and methods

2.1   General methods

The experiments proposed here are designed for *E. coli* but should be readily adaptable to other bacterial species for which there exist techniques for linear DNA transformation and phage transduction.

An *E. coli* reference strain and cloning vector are used. A *recD* (or *recBC sbcB*) mutant is required for the linear transformation step [9, 10]. A general transducing phage like P1*vir* can be used for transferring duplications from strain to strain. A selectable cassette is needed as well, e. g., a plasmid carrying an antibiotic resistance gene flanked by appropriate restriction sites.

Standard culture media, rich and defined, liquid and solid, and standard buffers are used throughout [11, 12]. Antibiotics, restriction enzymes, and DNA ligase are required as well.

Equipment for thermostatted cell growth, DNA extraction, purification, and digestion, and for transformation are needed.

All methods are standard: bacterial culture, P1 transduction, transformation, chromosomal and plasmid DNA extraction, DNA restriction and ligation, PCR amplification, Southern blots [11, 12].

2.2 Protocols

Protocol 1   Construction of precise tandem duplications on the chromosome

The commonest techniques for creating precise chromosomal deletions or deletion-insertions usually involve constructing the desired structure on a plasmid, then substituting this construct for the wild-type chromosomal region by two reciprocal recombination events. It is worthwhile to examine this familiar procedure in more detail, as parts of it are adapted in the technique for constructing precise duplications. If we represent the normal chromosomal sequence by the alphabet, a deletion of EFGHI, for example, will create a unique new sequence, the join CDJK (Fig. 1) . This join defines the deletion as extending precisely from the right of D to the left of J. One often chooses to replace the deleted sequence with a new sequence that provides a selection, such as an antibiotic resistance cassette. If we call this substitution $\alpha\beta\gamma$, the desired new sequence will be ABCD$\alpha\beta\gamma$JKL... The usual procedure is to construct the structure CD$\alpha\beta\gamma$JK (or CDJK) on a plasmid, generally in several steps. The sequences CD and JK provide homology for recombination with the chromosome. The two required recombinations are often carried out sequentially, with a selection for each. For example, one can select first for integration of a non-replicative plasmid then for loss of a conditionally harmful plasmid gene; loss of the integrated plasmid will either restore the wild-type sequence or replace it with the deletion-insertion (Fig. 2).

If one were to use the same technique to create a duplication of EFGHI, it would require constructing the sequence EFGHIEFGHI, or EFGHI$\alpha\beta\gamma$EFGHI if a selectable marker is to be associated with the duplication (Fig. 2). This entails several potential difficulties. One problem arises when the sequence to be duplicated is long. If, for example, EFGHI is 20 kb, the desired structure,

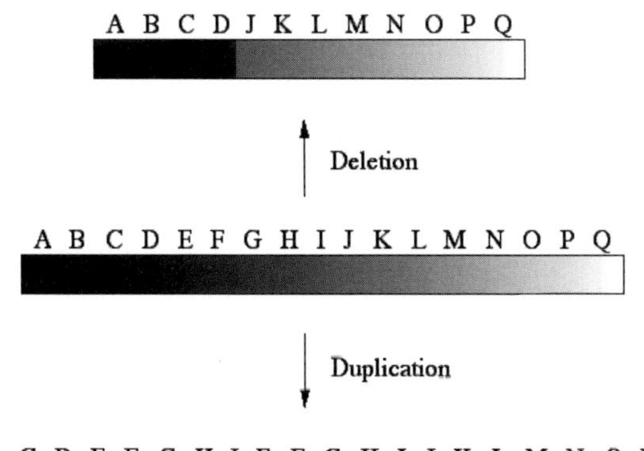

Figure 1   Unique new sequence created by deletions and duplications. The normal bacterial chromosome is shown (middle) along with a deleted chromosome (top) and a chromosome carrying a tandem duplication (bottom).

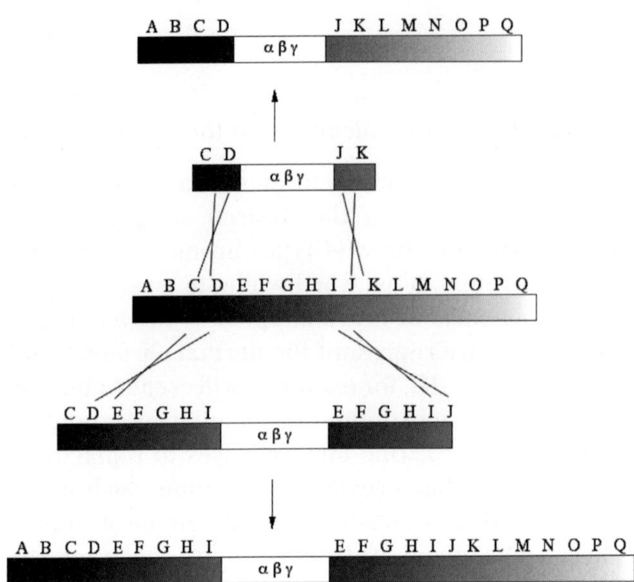

Figure 2 Classical technique to introduce deletions and duplications into the bacterial chromosome. Above and below the normal bacterial chromosome are schematized the *in vitro* constructs used to introduce a deletion or a duplication into the chromosome by linear transformation. αβγ represents the selectable marker, usually conferring antibiotic resistance.

exceeding 40 kb, could not be readily cloned in the usual plasmid vectors. A second major technical problem is the inherent instability of tandem duplications in plasmids.

These problems are avoided in our technique. Let us first note that a duplication, like a deletion, creates a unique new sequence (Fig. 1). If EFGHI is duplicated, this new sequence is the join HIEF, and, as in the case of a deletion, this sequence defines the precise extent of the duplication, i. e., from the left of E to the right of I. Cloning HIEF (or HIαβγEF if there is an associated cassette) poses no particular problem. As above, it can be done piecemeal in several easy steps. The resulting plasmid will have all the necessary information for defining the desired duplication without actually carrying the duplication itself. In fact, it would not even carry one copy of the full sequence to be duplicated; in our example, the sequence corresponding to G (which could represent 20 kb or more) is not present. A duplication of 20 kb or more thus can be precisely defined by a very short sequence constructed on a plasmid.

To construct a precise duplication of the sequence EFGHI, for example, the simplest procedure is to clone the various pieces sequentially in an appropriate vector. The chromosomal sequences EF and HI can be amplified by PCR, adding appropriate restriction sites at the ends, according to the cloning vector and cassette chosen. Care must be taken to clone the right-hand chromosomal fragment (HI) on the left in the plasmid construct and the left-hand chromosomal fragment (EF) on the right, maintaining the chromosomal orientation of both fragments. The cassette αβγ can then be cloned at the junction between HI and EF.

This construct, which, as we have seen, contains all the necessary information to determine the duplication endpoints, must now be introduced onto the chromosome. This turns out to be relatively straightforward. It requires two copies of the wild-type chromosomal region and three reciprocal recombination events. (In comparison, introduction of a deletion requires a single chromosomal copy and two recombinations.) The two copies can be on different chromosomes or on the replicated portion of a single structure. In our work [8], we used cells grown in rich medium (LB broth), favouring multiple chromosomes. The three homologous recombinations are illustrated in Figure 3. Somewhat surprisingly, when the duplication is associated with a selectable cassette, the efficiency of recombination is high enough to allow direct selection for the final product, rather than designing individual selections for each event. We found in transduction experiments (*cf.* section 3.2 below) that it was sufficient to introduce the construct HIαβγEF into the cell and select for the αβγ marker.

In practise, the cloned HIαβγEF sequence can be introduced into a *recD* (or *recBC sbcB*) mutant by linear transformation. The plasmid can be linearised by digestion with appropriate restriction enzymes, and the fragment containing the sequence HIαβγEF should be purified on an agarose gel. The selection in the transformation will be for the αβγ cassette marker. It must be checked that the transformants obtained have not received the selective marker of the cloning vector; these could arise from traces of contaminating uncut plasmid DNA. The presence of the duplicated sequence EFGHIαβγEFGHI can then be verified by PCR amplification of the new joins GHIα and γEFG. Furthermore, the transformants should exhibit slight instability: by reciprocal recombination, cultures will segregate approximately 1% of cells of the parental genotype, lacking the αβγ cassette and the duplication. We have shown [8] that the duplications are stabilised by the introduction of a *recA* mutation, which prevents homologous recombination.

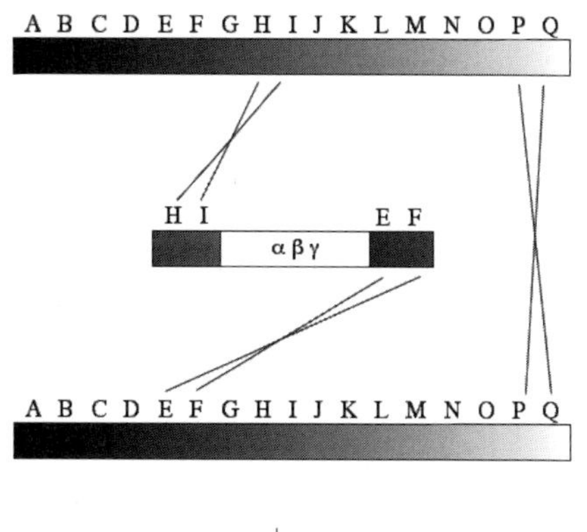

Figure 3 New way to create a duplication on the bacterial chromosome. Two chromosomes of the recipient cell are represented with, in between, the *in vitro* construct carrying the new sequence.

Protocol 2   Transfer of duplications from strain to strain

To transfer the duplication to other strains, linear transformation can be used again. Transformants can then be transduced to $recD^+$, if desired, by using a linked transposon insertion. Alternatively – and this is a tremendous technical advantage – the duplication can be transferred from strain to strain by P1-mediated transduction, again selecting for the αβγ marker. The transduction will work even if the entire duplication is too long to be packaged in a P1 phage head. As we have seen, the phage need only introduce the sequence HIαβγEF into the recipient. This contains all the information needed to reconstruct the same duplication, which is readily done by three recombination events. Using P1*vir*, which encapsidates about 90 kb, we routinely transduced duplications exceeding 200 kb in length [8].

Protocol 3   Construction of heterozygous partial diploids

It should be noted that when the duplication is transferred to a new strain, it is the recipient sequence that is duplicated. If the recipient carries a mutation G', for example, the transformants or transductants will have the structure ...CDEFG'HIαβγEFG'HIJ... [8]. Apart from the αβγ cassette, the only donor markers that can appear in the transformants or transductants are alleles tightly linked to α or γ, i. e., mutations in the regions I or E. In this case, the new strain will be heterozygous for that locus, with one of the duplicated sequences carrying the recipient allele and the other carrying the donor allele.

Although transformants and transductants are generally homozygous for the recipient alleles, heterozygous strains can be constructed quite easily, even if the target gene is not tightly linked to the αβγ cassette. Starting with a homozygous duplication of the sequence EFGHI, for example, one can introduce any selectable marker, e. g. G', by P1 transduction. If selection for αβγ is maintained, transductants selected for G' will acquire the allele by two reciprocal recombinations within one of the duplicated sequences (Fig. 4).

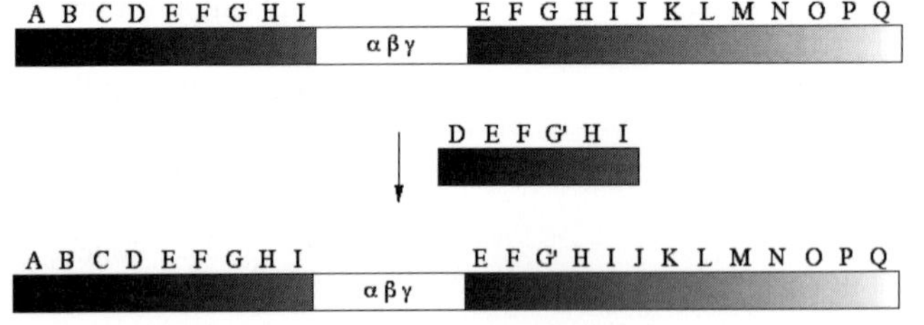

Figure 4   Construction of heterozygous diploids.

Strains with heterozygous duplicated regions, like homozygous partial diploids, exhibit slight instability. Segregants, by reciprocal recombination, lose the αβγ cassette and one copy of the duplication, therefore becoming haploid. In this case, though, there will be two types of haploid segregants: the original parental type (haploid for EFG'HI) and a haploid strain carrying the alternative allele (EFGHI).

3   Results and discussion

3.1   Further amplification of the duplicated region

It is possible, starting with a duplication, to amplify a sequence more than two-fold. This occurs spontaneously by unequal crossing-over between the two repeats, creating one haploid segregant and one triploid chromosome. The triploid can generate higher copy numbers, again by unequal crossing-over. In our own work, our selection for amplification of FtsQ, FtsA, and FtsZ required more than a doubling, and our mutants had more than two copies of the *ftsQAZ* region [8]. We were able to show triploidy for the nearby *leuABCD* operon by simultaneous maintenance of *leu⁺*, *leu*::Tn*9* and *leu*::Tn*10* alleles.

To produce a population with more than two copies of the duplicated region, the simplest approach is to include in the αβγ cassette a gene for which a higher expression level can be selected. The *ftsQAZ* operon could be used; growth in the presence of mecillinam would then select a population with higher amplification. A gene coding for β-lactamase could also be used; this enzyme destroys β-lactams like ampicillin, and the strain's level of ampicillin resistance is approximately proportional to the enzyme level over a fairly wide range. Other antibiotic resistance genes also can be used. Although the resistance level does not always increase linearly with gene copy number, it generally does increase. If a single copy already confers relatively high resistance, one can clone the gene under a weaker promoter.

3.2   Other amplifications

More elaborate amplification selections can be envisaged. For instance, one could take advantage of one of the many poison-antidote pairs known in bacteria. The gene coding for the poison would be cloned under control of an inducible promoter, and the antidote gene would be included in the αβγ cassette. Increased expression of the poison gene would then select for amplification of the antidote gene.

Multiploid strains with tandem repeats tend to be unstable and, in the absence of selective pressure to segregate cells with fewer copies. The number of copies can be stabilised by introduction of a *recA* mutation, which abolishes the unequal crossing-over needed for segregation. With a stabilised strain, the actual number of copies of the amplified region can be determined by quantitative PCR or Southern blots.

It will be readily appreciated that multiple tandem copies of a precise sequence can no longer be transferred from strain to strain by simple transduction unless, exceptionally, the entire amplified region can be encapsidated in the transducing phage, together with sufficient flanking DNA to permit recombination. In principle, amplified sequences in an Hfr strain could be crossed into F-recipients, although of course the latter would have to be *recA+*, at least temporarily, which could result in a change in the number of copies of the duplicated sequence. As a general rule, however, it would probably be advisable to set up a convenient selection scheme for amplification and then introduce the duplication into the various strains needed and select for further amplification in each strain separately.

4   Concluding remarks

The techniques of molecular biology have gradually introduced refinements in the handling of bacterial chromosomes. Instead of screening a large number of random mutations for those of interest, one can now use directed mutagenesis to introduce directly the substitution or frame mutation desired. Similarly, it is relatively easy to construct deletions and inversions with precisely defined endpoints, rather than collecting and characterising rare spontaneous deletions. Spontaneous duplications occur, generally by unequal crossing-over between nearly homologous natural sequences [13]. The present article offers the possibility of constructing precise duplications with virtually any endpoints and, from these, further multiples of a given DNA sequence.

References

1   Khattar MM (1997) Overexpression of the hslVU operon suppresses SOS-mediated inhibition of cell division in *Escherichia coli. FEBS Lett* 414: 402–404

2   D'Ari R, Casadesús J (1998) Underground metabolism. *BioEssays* 20: 181–186

3   Danese PN, Murphy CK, Silhavy TJ (1995) Multicopy suppression of cold-sensitive *sec* mutations in *Escherichia coli. J Bacteriol* 177: 4969–4973

4   Gordon GS, Sitnikov D, Webb CD et al. (1997) Chromosome and low copy plasmid segregation in *E. coli*: visual evidence for distinct mechanisms. *Cell* 90: 1113–1121

5   Tomic-Canic M, Bernerd F, Blumenberg M (1996) A simple method to introduce internal deletions or mutations into any position of a target DNA sequence. *Methods Mol Biol* 57: 249–257

6   Vinella D, Joseleau-Petit D, Thévenet D et al. (1993) Penicillin-binding protein 2 inactivation in *Escherichia coli* results in cell division inhibition, relieved by FtsZ overexpression. *J Bacteriol* 175: 6704–6710

7   Navarro F, Robin A, D'Ari R et al. (1998) Analysis of the effect of ppGpp on the *ftsQAZ* operon in *Escherichia coli. Mol Microbiol* 29: 815–823

8   Vinella D, Cashel M, D'Ari R (2000) Selected amplification of the cell division genes *ftsQ-ftsA-ftsZ* in *Escherichia coli. Genetics* 156: 1483–1492

9   Russell CB, Thaler DS, Dahlquist FW (1989) Chromosomal transformation of *Escherichia coli recD* strains with linear-

ized plasmids. *J Bacteriol* 171: 2609–2613

10 Winans SC, Elledge SJ, Krueger JH et al. (1985) Site-directed insertion and deletion mutagenesis with cloned fragments in *Escherichia coli. J Bacteriol* 161: 1219–1221

11 Miller JH (1992) *A short course in bacterial genetics.* Cold Spring Harbor Laboratory Press, Plainview, NY

12 Sambrook J, Fritsch EF, Maniatis T (1989) *Molecular cloning: a laboratory manual.* Cold Spring Harbor Laboratory, Cold Spring Harbor, N. Y.

13 Anderson P, Roth J (1981) Spontaneous tandem genetic duplications in *Salmonella typhimurium* arise by unequal recombination between rRNA (*rrn*) cistrons. *Proc Natl Acad Sci USA* 78: 3113–3117

Generalized Transduction

Anne Thierauf and Stanley Maloy

Contents

1   Introduction

Transduction is a phenomenon in which bacterial DNA is transferred from one bacterial cell to another by a phage particle. There are two types of transduction: generalized transduction and specialized transduction.

Generalized transducing phage are produced by an error during packaging of DNA into a phage head, resulting in the insertion of random, phage-size fragments of bacterial DNA into the phage head instead of phage DNA. Thus, in a lysate of generalized transducing phage, some particles contain DNA obtained from the host cell rather than phage DNA. The bacterial DNA fragment packaged within a phage head can be derived from any part of the host genome, although certain regions of the genome may be packaged at varying frequencies. Phage particles that contain host DNA are called transducing particles. When a transducing particle adsorbs to a sensitive recipient, the double-stranded DNA is injected into the recipient cell. Stable inheritance of the donor DNA requires either that it integrates into the recipient chromosome *via* homologous recombination or that it is replicated (e.g., following transfer of

Methods and Tools in Biosciences and Medicine
Prokaryotic Genomics, ed. by M. Blot
© 2003 Birkhäuser Verlag Basel/Switzerland

plasmids). Generalized transducing particles can be produced during lytic growth of either virulent or lysogenic phage. For detailed reviews on generalized transduction, see Margolin [1] and Masters [2, 3].

Specialized transducing phage result from the aberrant excision of an integrated lysogenic phage. In contrast to generalized transducing phage, only regions of the host DNA that flank the prophage are packaged in specialized transducing phage, allowing transduction of a contiguous DNA fragment with both host and phage DNA. Because it is possible to use genetic tricks to coax lysogenic phage to integrate at a variety of sites in the genome, specialized transducing phage can be isolated from many different regions of a bacterial genome. Stable inheritance of specialized transducing fragments may occur either by site-specific recombination mediated by the phage integrase or by homologous recombination mediated by bacterial recombinases.

2   Materials

2.1   Bacterial strains and phage

Salmonella strains and P22 HT *int* phage can be obtained from the *Salmonella* Genetic Stock Centre at the University of Calgary (www.ucalgary.ca/~kesander/). P22 HT *int* has a turbid plaque phenotype. Rare clear plaque mutants that reproduce more rapidly than the turbid plaque phage can accumulate in a P22 HT *int* lysate after it is repeatedly propagated. Such clear plaque mutants rapidly lyse infected cells, decreasing the number of transductants obtained. Therefore, P22 HT *int* phage stocks should be occasionally checked for clear plaque mutants; if this is a problem, a new lysate should be prepared from a single isolated turbid plaque.

E. coli strains and P1 phage can be obtained from the *E. coli* Genetic Stock Center at Yale University (cgsc.biology.yale.edu/).

2.2   Reagents, media, and solutions

LB medium (also known as Lysogeny Broth or Luria-Bertani broth), EBU medium, TS top agar, and P22 broth can be prepared as described in Maloy et al. [4]. Salt solutions should be prepared in deionized H_2O.

3   Methods

Transduction requires three steps: growth of a phage lysate on the donor strain, infection of a recipient strain, and selection of transductants. It is often necessary to avoid co-infection with viable phage. The methods used depend upon the properties of the specific transducing phage.

Protocol 1   P22 transduction (modified from Maloy, [5])

Preparation of P22 HT int phage lysates

1. Pick a single colony and start an overnight culture of the donor strain in 1 ml LB at 37 °C.
2. Add 1 ml P22 broth to 200 μl of the overnight culture. The final multiplicity of infection (MOI) should be about 0.01–0.1 pfu/cell.
3. Incubate 8–16 h in a 37 °C shaker. (Temperature sensitive mutants can be grown at 30 °C.)
4. Add several drops of chloroform and vortex. Transfer the supernatant to a microfuge tube and centrifuge for 1 min at 12,000 rpm to pellet the cell debris.
5. Transfer the supernatant to a new microfuge tube. Add several drops of chloroform and vortex. Store at 4 °C. (A good lysate should contain 10^{10}–10^{11} pfu/ml.)

P22 transduction without phenotypic expression

1. Grow the recipient strain overnight in 1 ml LB.
2. Dilute the phage by adding 50 μl of the P22 HT *int* lysate to 450 μl 0.85% NaCl. Mix the cells and freshly diluted phage lysate as follows:

Plate	ml cells	ml P22	
A	200 μl	–	No cell control
B	–	50 μl	No phage control
C	200 μl	10 μl	
D	200 μl	50 μl	
E	200 μl	200 μl	

3. Sterilize a glass spreader by dipping it into alcohol then briefly passing it through a flame to burn off the residual alcohol. Thoroughly spread the cells and phage on the plates, re-sterilizing the glass spreader between each plate.
4. Place the plates in an incubator upside down. Incubate 1–2 days at 37 °C. (Temperature-sensitive mutants can be grown at 30 °C.)

5. Count the colonies on each plate. There should be no growth on the cell or phage control plates. Any transductants that will be saved should be immediately purified on EBU plates and cross-streaked against phage P22-H5 to purify phage-sensitive colonies.

P22 transduction with phenotypic expression

For certain transductions (e.g., when selecting Kan^R or Str^R), phenotypic expression is required before plating on the selective medium. Phenotypic expression can be done in two ways:
- Spread the cells and phage on nonselective medium (e.g., LB plate), incubate 4–8 h, then replica plate onto the selective medium. This replica plating approach usually gives more colonies and ensures that different colonies are not due to siblings.
- Mix cells and phage in a microfuge tube and incubate about 1 h at 37 °C before plating on the selective medium. Although this broth approach may yield somewhat fewer colonies and may produce some siblings, it is easier and faster than the replica plating approach. The broth method is detailed below.

1. Perform steps 1–2 as described above.
2. Leave at room temperature for 15–30 min to allow phage adsorption.
3. Add 1 ml LB to each tube. Incubate at 37 °C for 1 h to allow phenotypic expression.
4. Centrifuge each tube for 1 min in a microfuge.
5. Pour off the supernatant. Add 100 μl LB, and vortex to resuspend the pellet.
6. Continue with steps 3–5 as described above.

Spot titering P22 lysates

When P22 HT *int* is used, phage stocks usually do not need to be titered. However, it is a good idea to check the titer of your phage stock if a transduction does not work.

1. Mark 4 quadrants on the bottom of an EBU plate. The surface of the plate should not be noticeably wet. If the plate is wet, place it in an incubator or oven with the lid ajar until dry.
2. Melt TS top agar in a microwave. For each phage to be titered, add 2.5 ml of the melted top agar to a test tube and place in a 50 °C heating block. After the top agar cools to about 50 °C, add 0.1 ml of an overnight culture of a P22-sensitive *Salmonella* strain to each tube.
3. Immediately swirl and pour onto the EBU plate.
4. Allow the top agar to solidify for 15–30 min.
5. Gently spot 20 μl of appropriate phage dilutions onto quadrants on the plate (usually 10^{-6}, 10^{-7}, 10^{-8}, 10^{-9} dilutions in sterile 0.85% NaCl).
6. Leave the plate on the bench for about 30 min or until the drops of phage dry.
7. Incubate the plates upside down at 30–37 °C overnight.

8. Confluent growth of the bacteria in the top agar will result in a "lawn" and plaques will appear where phage has been spotted. Count the number of plaques in each spot. Each viable phage (plaque-forming unit or pfu) will produce one plaque. Calculate the phage titer as follows:

$$pfu/ml = \frac{number\ of\ plaques \times dilution\ factor}{20\mu l} \times \frac{1000\mu l}{ml}$$

Checking Salmonella *for P22 sensitivity*

In addition to receiving a transducing fragment, some of the transductants also may have been infected with P22 phage. Although P22 HT *int* is unable to integrate into the chromosome because of the lack of integrase, the phage can form unstable, episomal pseudolysogens. Pseudolysogens can be differentiated from nonlysogens and true lysogens on EBU plates [6]. EBU plates contain pH indicators, which are green at neutral pH but turn dark blue at low pH. When streaked on green plates, nonlysogens and true lysogens form light-colored colonies. However, in a colony containing pseudolysogens, many cells are undergoing lysis, which lowers the pH of the medium and results in dark blue colonies. The phage DNA in pseudolysogens is replicated as a low copy number plasmid, so it is possible to obtain "phage-free" segregants by simply streaking for isolated colonies, on EBU plates. The "phage-free" segregants will form light-colored colonies, while the pseudolysogens will remain blue.

P22 HT *int* is used for transductions because the *int* mutation prevents formation of stable lysogens. However, when cells are left on plates with lytic phage, there is a strong selection for revertants that form stable lysogens. Since stable lysogens cannot be reinfected with P22, such transductants are not very useful for genetic studies. Therefore, it is important to isolate "phage-free" transductants as soon as possible after colonies arise. Ethylene glycol-bis[β-aminoether] N,N,N′,N′-tetra-acetic acid (EGTA) can be included in plates to chelate the Ca^{+2} required for phage adsorption, thereby preventing reinfection of transductants [7]. However, EGTA cannot be added during the initial phage infection, or it will prevent adsorption of the transducing phage. Even colonies from medium containing EGTA must be cleaned up on EBU plates. Once light-colored colonies have been purified from EBU plates, they should be checked to make sure they are not true lysogens. This is done by cross-streaking against a P22 *c2* mutant called H5. The P22 *c2* gene encodes a repressor equivalent to cI of phage lambda. Phage-free cells are infected by P22-H5 and lysed, but P22 lysogens are resistant to P22-H5.

Avoid "digging into" the agar when streaking EBU plates, because when growing anaerobically, the bacteria ferment more glucose and all of the colonies will appear dark blue. Also, the EBU phenotype should be observed promptly after growth appears, because when left on EBU plates for many days, all of the colonies will turn dark-colored.

Protocol 2   Transduction with P1*vir* (modified from Silhavy et al. [8])

Preparation of P1 phage lysates

1. Pick a single colony and start an overnight culture of the donor strain in 1 ml LB. Grow with aeration at 37 °C.
2. Subculture 10 µl of the overnight culture into 1 ml LB containing 0.2% glucose and 5 mM $CaCl_2$.
3. Incubate for 30 min with aeration at 37 °C. An early exponential phase culture is optimal for preparation of a high-titer P1 lysate.
4. Add 10 µl of a P1 lysate with approximately 5×10^8 P1*vir* per ml.
5. Incubate with aeration at 37 °C until the culture lyses or for 2–3 h. Typically, the cells will lyse during this time, but a usable phage titer may be obtained even if the culture does not clear.
6. Add 0.1 ml chloroform and vortex.
7. Transfer the supernatant to a sterile microfuge tube and centrifuge at 12,000 rpm for 1 min to pellet the debris.
8. Transfer the supernatant to a sterile tube. Add 0.1 ml chloroform, vortex, and store at 4 °C.

P1 transduction without phenotypic expression

1. Pick a single colony and start an overnight culture of the recipient strain in 1 ml LB. Grow with aeration at 37 °C.
2. Centrifuge the overnight culture for 30 s at 12,000 rpm in a microfuge.
3. Resuspend the cell pellet in 0.5 ml of CaMg (5 mM $CaCl_2$ + 10 mM $MgSO_4$).
4. Mix the cells and phage in sterile microfuge tubes as follows:

Plate	ml cells	ml P1	
A	100 µl	–	No cell control
B	–	100 µl	No phage control
C	100 µl	10 µl	
D	100 µl	50 µl	
E	100 µl	100 µl	

5. Leave at room temperature for about 30 min to allow phage adsorption.
6. Add 100 µl 1 M sodium citrate to each tube and vortex briefly. (The citrate chelates the divalent cations, preventing subsequent re-infection.)
7. Pipet the contents of each tube to a separate plate of agar medium formulated to select for the desired transductants.
8. Dip a glass hockey-stick into alcohol then briefly pass it through a flame to burn off the residual alcohol. Thoroughly spread the plates with the alcohol-flamed glass spreader.
9. Place the plates in an incubator upside down. Incubate overnight at 30–37 °C.
10. Count the colonies on each plate. There should be no growth on the cell or phage control plates.

P1 transduction with phenotypic expression

1. Perform steps 1–5 as described above.
2. Add 1 ml LB to each tube. Incubate at 37 °C for 1 h to allow phenotypic expression.
3. Centrifuge for 30 s at 12,000 rpm in a microfuge to pellet the cells.
4. Pour off the supernatant. Resuspend the cells in 0.1 ml LB containing 20 mM sodium citrate.
5. Continue with steps 7–9 as described above.

Spot titering P1 lysates

Because using several different MOIs with a P1 sensitive *E. coli* recipient typically yields transductants, titering phage lysates is usually unnecessary. However, it may be useful to titer a phage stock if the transduction fails. Although LB plates will work for this procedure, P1 plaques are small and it is easier to count plaques on EBU plates because the plaques produce dark blue spots on a green background.

1. Prepare an overnight culture of a sensitive *E. coli* strain in 1 ml LB at 37 °C.
2. Centrifuge the cells for 30 s at 12,000 rpm in a microfuge.
3. Resuspend the cell pellet in 0.5 ml 10 mM $MgCl_2$. Store on ice until use.
4. Mark 4 quadrants on the bottom of an EBU plate. The surface of the plate should not be noticeably wet. If the plate is wet, place it in an incubator or oven with the lid ajar until dry.
5. Melt TS top agar in a microwave. For each phage to be titered, add 2.5 ml of the melted top agar to a test tube and place in a 50 °C heating block. After the top agar cools to about 50 °C, add 0.1 ml of the *E. coli* suspension in 10 mM $MgCl_2$ to each tube.
6. Immediately swirl and pour onto the EBU plate. Allow the top agar to solidify for 15–30 min.
7. Gently spot 20 µl of appropriate phage dilutions onto quadrants of the plate (usually 10^{-6}, 10^{-7}, 10^{-8}, 10^{-9} dilutions in sterile 0.85% NaCl).
8. Leave the plate on the bench for about 30 min or until the drops of phage dry.
9. Incubate the plates upside down at 30–37 °C overnight.
10. Confluent growth of the bacteria in the top agar will result in a "lawn" and plaques will appear where phage has been spotted. Count the number of plaques in each spot. Each viable phage (plaque forming unit or pfu) will produce one plaque. Calculate the phage titer as follows:

$$pfu/ml = \frac{number\ of\ plaques \times dilution\ factor}{20\mu l} \times \frac{1000\mu l}{ml}$$

4   Results and discussion

4.1   Generalized transduction by phage P22

Phage P22 is the most widely used transducing phage in *Salmonella*. P22 is a temperate phage that adsorbs to *Salmonella* by binding to the O-antigen of the lipopolysaccharide on the outer membrane. After infection, the linear double-stranded P22 genome circularizes by recombination between terminal redundancies at each end of the phage DNA. During lytic growth, the circular genome of P22 undergoes several rounds of bidirectional replication, then changes to rolling-circle replication. Rolling-circle replication produces long concatemers of double-stranded P22 DNA. These concatemers are packaged into phage heads by a "headful" mechanism. Packaging is initiated at a specific 8–10 bp sequence on the DNA called a *pac* site. A phage-encoded endonuclease initially cuts the DNA at the *pac* site and initiates packaging of the DNA into a phage head. One phage head is filled with 44 Kb of DNA. Then the concatameric DNA protruding from the base of the phage head is cut by the endonuclease and transferred to an empty phage head. These reactions continue processively until 3–5 phage heads are filled with DNA [9]. The biology of phage P22 is reviewed in Susskind and Botstein [10] and Poteete [11].

When the host cell lyses, it releases 50–100 new phage. The released phage infect other cells in the culture, and after many rounds of phage multiplication and lysis of the bacterial culture, the broth contains a high concentration of phage (about 10^{10}–10^{11} pfu/ml). Chloroform is added to kill any unlysed cells, and cell debris is removed by centrifugation, yielding a solution of phage called a "phage lysate" or "phage stock". If cells or cell debris remain in a lysate, the phage will adsorb to them, decreasing the effective titer of phage. P22 lysates can be stored for many years at 4 °C or frozen in 2.5 M glycerol at –70 °C.

Occasionally, host DNA is packaged into P22 heads as a result of the erroneous recognition of sequences in the *Salmonella* genome that are homologous to the P22 *pac* site. When P22 infects a cell, occasionally the P22 nuclease cuts one of these chromosomal pseudo-*pac* sites and initiates processive packaging of 44 Kb chromosomal DNA fragments into P22 phage heads [1, 2]. Because pseudo-*pac* sites are distributed unevenly around the *Salmonella* chromosome, the frequency of transduction by wild-type P22 varies markedly for different regions of the chromosome. The P22 particles carrying chromosomal DNA (transducing particles) can inject this DNA into a new host.

A derivative of phage P22, P22 HT105/1 *int-201*, is typically used for generalized transduction [12]. This phage has a mutant nuclease with less specificity for the *pac* sequence, resulting in essentially random packaging of chromosomal DNA and a high transducing (HT) frequency [13]. About 50% of the P22 HT phage heads carry random transducing fragments of chromosomal DNA [14, 15]. The *int* mutation decreases formation of stable lysogens that would interfere with subsequent transductions.

4.2   Generalized transduction by phage P1

Phage P1 is the most commonly used transducing phage in *E. coli*. P1 is a temperate phage that infects a variety of a gram-negative bacteria by binding to the terminal glucose residues of the lipopolysaccharide core on the outer membrane. Interaction of P1 tail fibers with the receptor requires Ca^{+2}. After infection, the linear double-stranded P1 DNA circularizes by recombination. During lytic growth, the circular genome of P1 undergoes several rounds of bidirectional replication, then switches to rolling circle replication, which yields long concatemers of double-stranded P1 DNA. Packaging is initiated at a specific 162 bp *pac* sequence on the phage DNA. A phage encoded endonuclease cuts the DNA and facilitates headful packaging into a phage head. Each phage head accommodates about 110–115 Kb of DNA. After the head is filled, the excess DNA is cleaved by a sequence independent mechanism. Subsequent rounds of packaging are then initiated from the cleaved DNA. These reactions continue processively until 3–5 phage heads are filled with DNA. When a cell lyses, it releases 25–150 new phage. A typical P1 lysate contains about 10^9 pfu/ ml of phage. The biology of phage P1 is reviewed in Yarmolinsky and Sternberg [16].

P1 transducing particles carry DNA from different regions of the *E. coli* chromosome at about equal frequencies, implying that P1 transducing phage are not generated by packaging of the host chromosome at pseudo-*pac* sites. Packaging of the host DNA probably initiates at random breaks that arise during the infection [7].

In contrast to P22, phage P1 exists as an extrachromosomal plasmid during lysogeny. P1 lysogens express a Type III restriction system that can interfer with subsequent transductions. Lysogens can be avoided by using a derivative called P1vir[s] that constitutively produces a phage antirepressor that interferes with the c1 repressor function of phage P1. Other derivatives of P1 are available that use alternative ploys to avoid stable lysogeny.

4.3   Generalized transduction in other bacteria

Other phage also have been isolated that allow generalized transduction in *Salmonella* and *E. coli*, and specific generalized transducing phage have been isolated for a variety of other bacteria (Tab. 1). In addition, mutations that alter the host-range of phage by allowing binding to a different host receptor can expand the host-range for generalized transducing phages [3]. Although some phage yield only low frequencies of generalized transduction, it may be possible to isolate higher transducing derivatives by using the approach described by Schmieger [14].

Table 1 Examples of generalized transducing phage

Bacteria	Phage	Reference
Acetobacter methanolicus	Acm1	[21]
Acinetobacter		[22]
Bordetella avium	Ba1	[23]
Bacillus stearothermophilus	TP-42, TP-56 (temperate) TP-68 (virulent)	[24]
Bacillus megaterium	MP13	[25]
Bacillus thuringiensis, B. cereus, B. anthracis	CP-51, CP-54	[26]
Caulobacter crescentus	phi Cr30	[27]
Escherichia coli	P1, phi w39, T1, T4, Mu	[3, 28, 29]
Listeria monocytogenes	P35 (phiLMUP35), U153(phiCU-SI153/95)	[30]
Myxococcus xanthus	MX4	[31, 32]
Proteus mirabilis	phim and pi1	[33]
Proteus mirabilis, P. vulgaris	5006MHFTk	[34]
Pseudomonas aeruginosa	UT1, B86	[35, 36]
Pseudomonas cepacia	CP75	[37]
Rhizobium meliloti	phage 11	[38]
Salmonella enterica	P22, SE1, KB1, L, ES18	[3, 20, 39–41]
Salmonella enterica sv. Typhi	j2	[42, 43]
Serratia marcescens	3M	[44, 45]
Serpulina hyodysenteriae	VSH-1	[46]
Staphylococcus aureus	phage 80 alpha	[47]
Streptomyces coelicolor, S. avermitilis, S. verticillus	DAH2, DAH4, DAH5, DAH6	[48]
Streptomyces venezuelae	SV1	[49, 50]
Streptomyces hygroscopicus	SH10	[51]
Vibrio cholerae	CP-T1	[52]
Vibrio sp. strain 60	As3	[53]
Vibrio parahaemolyticus, V. alginolyticus	phi VP253, phi VP143	[54]
Xanthomonas campestris	XTP1	[55]
Methanobacterium thermoautotrophicum	psi M1	[56]

4.4 Inheritance of chromosomal DNA via homologous recombination

Genetic recombination involves the physical breakage, exchange, and rejoining of two DNA molecules. Homologous recombination can be mediated by several different pathways in bacteria, but inheritance of double-stranded, linear DNA *via* transduction primarily occurs *via* the RecBCD pathway. RecBCD initiates processing of a DNA fragment from a free double-stranded end and facilitates loading of RecA protein onto the DNA. RecA aligns homologous DNA molecules, promoting genetic exchange.

Homologous recombination between the chromosome and the linear, double-stranded DNA donated by generalized transduction typically occurs soon after the DNA enters a bacterial recipient. Upon infection with a transducing particle, only about 10% of the DNA fragments stably recombine with the chromosome. In the other recipients, a phage-encoded protein binds to the ends of the DNA fragment [17], and because this DNA lacks a free end, it is not an efficient substrate for homologous recombination. Such "abortive transductants" are rapidly segregated from the population during cell division.

Generalized transduction is an efficient method to move mutations between strains to construct derivatives with different genotypes. In order to determine the genetic or biochemical effects of a particular mutation, it is necessary to compare the mutant with a strain that only differs by a single mutation (an "isogenic strain"). If other uncharacterized mutations are present, it is not possible to determine whether the mutation of interest is responsible for the observed phenotypic change. The most common way to ensure that two strains are isogenic is to transfer a small region of DNA carrying the mutation into the parental strain by recombination. If a mutation confers a selectable phenotype, it can be transferred easily from a donor to an appropriate recipient strain. If the mutation cannot be selected directly, linkage to an adjacent, selectable marker can be used to move the mutation into a recipient strain. For example, an auxotrophic mutation may be brought into a recipient strain by selecting for inheritance of a nearby gene then screening for co-inheritance of the auxotrophic mutation.

Transduction also can be used to determine the relative map location of genes. Because the probability of a rare recombination event between any two adjacent base-pairs within long homologous DNA sequences is roughly equal, the physical distance separating two loci on the same DNA molecule determines the frequency of recombination in the intervening region. A recombination event between loci on the same DNA molecule will prevent their co-inheritance. Thus, if two loci are close to each other, they are more likely to be co-inherited than if the genes are farther apart. Genetic mapping exploits the frequency of co-inheritance to measure the relative distance between genetic loci. The frequency that two genes are co-inherited is defined as their linkage. The Wu formula provides an approximate correlation between cotransduction frequency and physical distance (Fig. 1). Determining the linkage of two genetic markers is called a two-factor cross. It is also possible to determine the relative location of genetic mutations by using three genetic markers (three-factor crosses) or by genetic crosses *versus* a set of defined deletion mutations (deletion mapping). Although it is also possible to determine the relative location of genes by DNA hybridization or DNA sequencing, genetic mapping often provides a simple and inexpensive way to rapidly determine the location of mutations in bacteria.

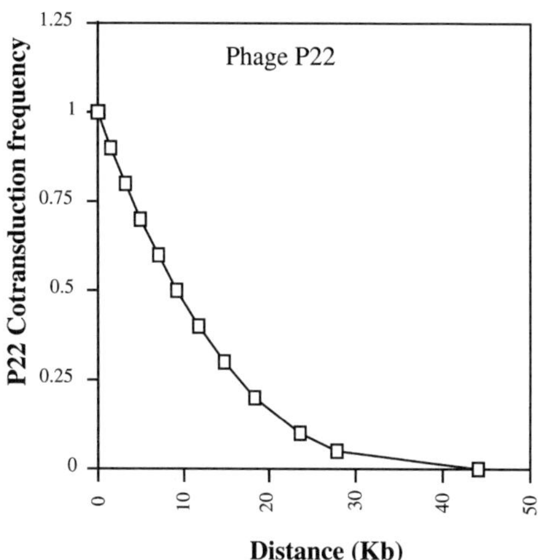

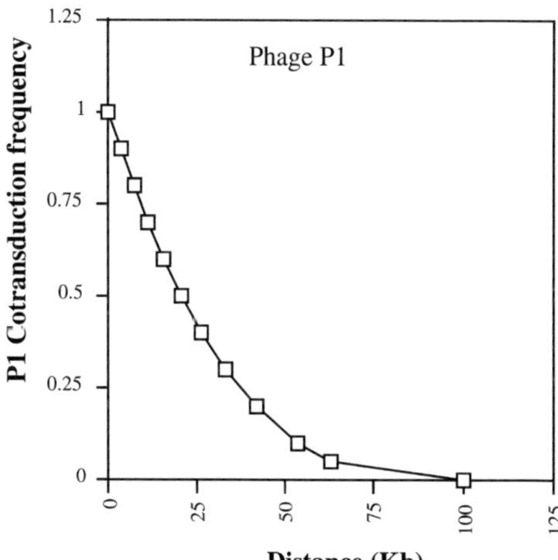

Figure 1 Cotransduction frequency vs distance calculated by the Wu formula. The Wu formula is based upon the relationship $d = L - (L \times \sqrt[3]{C})$, where d = distance between markers, L = length of phage DNA, and C = cotransduction frequency.

Transduction also provides a tool for isolation of new mutations *via* localized mutagenesis [4]. The donor bacteria can be treated with a mutagen before preparing a generalized transducing lysate, or a phage lysate can be directly treated with a mutagen. The mutagenized lysate is then transferred into recipient bacteria, selecting for a dominant genetic marker and screening for co-inheritance of mutations in a linked gene.

4.5 Transduction of plasmids

In addition to chromosomal genes, generalized transduction also can be used for transfer of plasmid DNA. A variety of phage can transfer plasmid DNA, but P22 has been exploited most effectively for transduction of plasmids between *Salmonella* strains. One mechanism for transfer of high copy number plasmid relies upon phage gene products that inhibit the host RecBCD exonuclease functions. This allows plasmids to replicate *via* rolling-circle replication, generating long plasmid concatemers that are substrates for packaging by P22 HT into phage heads [18]. A second mechanism requires that the plasmid carry some homology with the phage DNA [19]. Homologous recombination within the shared sequence results in integration of the plasmid into the phage genome and subsequent transfer of the plasmid into a recipient cell. A third mechanism for transfer of low copy number plasmids relies upon integration of the plasmid into the host chromosome, packaging of the fragment of host chromosome by P22 HT, and subsequent transfer of the fragment into a recipient cell [20]. Upon transduction into a recipient host *via* any of these three mechanisms, recombination between the tandemly repeated sequences results in re-circularization of the plasmid DNA. The circularized DNA can subsequently replicate as an autonomous plasmid.

Acknowledgments

We thank Jim Slauch for helpful discussions. This work was supported by grants from the NIH (PHS GM34715) and the USDA (AG 2001–35201–09950).

References

1   Margolin P (1987) Generalized transduction. In: F Neidhardt, J Ingraham, K Low, et al. (eds): *Escherichia coli and Salmonella typhimurium: Cellular and molecular biology.* American Society for Microbiology, Washington, DC

2   Masters M (1985) Generalized transduction. In: J Scaife, D Leach, A Galizzi (eds): *Genetics of bacteria.* Academic Press, New York, 197–205

3   Masters M (1996) Generalized transduction. In: FC Neidhardt (ed): *Escherichia coli and Salmonella: Cellular and molecular biology.* American Society for Microbiology, Washington, DC, 2421–2441

4   Maloy S, Stewart V, Taylor R (1996) *Genetic analysis of pathogenic bacteria.* Cold Spring Harbor Laboratory, Cold Spring Harbor, NY

5   Maloy S (1989) *Experimental techniques in bacterial genetics.* Jones and Bartlett, Boston, MA

6   Bochner B (1984) Curing bacterial cells of lysogenic viruses by using UCB indicator plates. *BioTechniques* 2: 234–240

7   Sternberg NL, Maurer R (1991) Bacteriophage-mediated generalized transduction in *Escherichia coli* and *Salmonella typhimurium. Methods Enzymol* 204: 18–43

8   Silhavy TJ, Berman ML, Enquist LW, et al. (1984) *Experiments with gene fusions.* Cold Spring Harbor Laboratory, Cold Spring Harbor, N.Y.

9   Casjens S, Hayden M (1988) Analysis in vivo of the bacteriophage P22 headful nuclease. *J Mol Biol* 199: 467–74

10 Susskind MM, Botstein D (1978) Molecular genetics of bacteriophage P22. *Microbiol Rev* 42: 385–413

11 Poteete AR (1988) Bacteriophage P22. In: R Calender (ed): *The bacteriophages*. Plenum Press, New York, 647–682

12 Davis R, Botstein D, Roth RJ (1980) *Advanced bacterial genetics*. Cold Spring Harbor Laboratory, Cold Spring Harbor, NY

13 Casjens S, Sampson L, Randall S, et al. (1992) Molecular genetic analysis of bacteriophage P22 gene 3 product, a protein involved in the initiation of headful DNA packaging. *J Mol Biol* 227: 1086–1099

14 Schmieger H (1972) Phage P22-mutants with increased or decreased transduction abilities. *Mol Gen Genet* 119: 75–88

15 Schmieger H, Backhaus H (1976) Altered cotransduction frequencies exhibited by HT-mutants of *Salmonella*-phage P22. *Mol Gen Genet* 143: 307–309

16 Yarmonlinsky MB, Sternberg N (1988) Bacteriophage P1. In: R Calendar (ed): *The bacteriophages*. Plenum Press, New York, 291–438

17 Benson NR, Roth J (1997) A *Salmonella* phage-P22 mutant defective in abortive transduction. *Genetics* 145: 17–27

18 Garzon A, Cano DA , Casadesus J (1995) Role of Erf recombinase in P22-mediated plasmid transduction. *Genetics* 140: 427–434

19 Orbach MJ, Jackson EN (1982) Transfer of chimeric plasmids among *Salmonella typhimurium* strains by P22 transduction. *J Bacteriol* 149: 985–994

20 Mann BA, Slauch JM (1997) Transduction of low-copy number plasmids by bacteriophage P22. *Genetics* 146: 447–456

21 Kiesel B, Wunsche L (1993) Phage Acm1-mediated transduction in the facultatively methanol- utilizing *Acetobacter methanolicus* MB 58/4. *J Gen Virol* 74: 1741–1745

22 Herman NJ, Juni E (1974) Isolation and characterization of a generalized transducing bacteriophage for *Acinetobacter*. *J Virol* 13: 46–52

23 Shelton CB, Crosslin DR, Casey JL, et al. (2000) Discovery, purification, and characterization of a temperate transducing bacteriophage for *Bordetella avium. J Bacteriol* 182: 6130–6136

24 Welker NE (1988) Transduction in *Bacillus stearothermophilus. J Bacteriol* 170: 3761–3764

25 Vary PS, Garbe JC, Franzen M, et al. (1982) MP13, a generalized transducing bacteriophage for *Bacillus megaterium. J Bacteriol* 149: 1112–1119

26 Thorne CB (1978) Transduction in *Bacillus thuringiensis. Appl Environ Microbiol* 35: 1109–1115

27 Bender RA (1981) Improved generalized transducing bacteriophage for *Caulobacter crescentus. J Bacteriol* 148: 734–735

28 Yoshida Y, Mise K (1984) Characterization of generalized transducing phage phi w39 heteroimmune to phage P1 in *Escherichia coli* W39. *Microbiol Immunol* 28: 415–426

29 Young KK, Edlin G (1983) Physical and genetical analysis of bacteriophage T4 generalized transduction. *Mol Gen Genet* 192: 241–246

30 Hodgson DA (2000) Generalized transduction of serotype 1/2 and serotype 4b strains of *Listeria monocytogenes. Mol Microbiol* 35: 312–323

31 Campos JM, Geisselsoder J, Zusman DR (1978) Isolation of bacteriophage MX4, a generalized transducing phage for *Myxococcus xanthus. J Mol Biol* 119: 167–178

32 Geisselsoder J, Campos JM, Zusman DR (1978) Physical characterization of bacteriophage MX4, a generalized transducing phage for *Myxococcus xanthus. J Mol Biol* 119: 179–189

33 Nakamura M, Horiuchi S, Nakaya R (1975) Comparative studies on generalized transducing bacteriophages of *Proteus mirabilis*, phim and pi1. *Jpn J Microbiol* 19: 123–131

34 Coetzee JN (1975) Transduction of a *Proteus vulgaris* strain by a *Proteus mirabilis* bacteriophage. *J Gen Microbiol* 89: 299–309

35 Kilbane JJ, Miller RV (1988) Molecular characterization of *Pseudomonas aeruginosa* bacteriophages: identification

and characterization of the novel virus B86. *Virology* 164: 193–200

36 Ripp S, Ogunseitan OA, Miller RV (1994) Transduction of a freshwater microbial community by a new *Pseudomonas aeruginosa* generalized transducing phage, UT1. *Mol Ecol* 3: 121–126

37 Matsumoto H, Itoh Y, Ohta S, et al. (1986) A generalized transducing phage of *Pseudomonas cepacia*. *J Gen Microbiol* 132: 2583–2586

38 Sik T, Horvath J, Chatterjee S (1980) Generalized transduction in *Rhizobium meliloti*. *Mol Gen Genet* 178: 511–516

39 Sander M, Schmieger H (2001) Method for host-independent detection of generalized transducing bacteriophages in natural habitats. *Appl Environ Microbiol* 67: 1490–1493

40 Schicklmaier P, Schmieger H (1995) Frequency of generalized transducing phages in natural isolates of the *Salmonella typhimurium* complex. *Appl Environ Microbiol* 61: 1637–1640

41 Llagostera M, Barbe J, Guerrero R (1986) Characterization of SE1, a new general transducing phage of *Salmonella typhimurium*. *J Gen Microbiol* 132: 1035–1041

42 Mise K, Kawai M, Yoshida Y, et al. (1981) Characterization of bacteriophage j2 of *Salmonella typhi* as a generalized transducing phage closely related to coliphage P1. *J Gen Microbiol* 126: 321–326

43 Mise K, Yoshida Y, Kawai M (1983) Generalized transduction between *Salmonella typhi* and *Salmonella typhimurium* by phage j2 and characterization of the j2 plasmid in *Escherichia coli*. *J Gen Microbiol* 129: 3395–3400

44 Regue M, Fabregat C, Vinas M (1991) A generalized transducing bacteriophage for *Serratia marcescens*. *Res Microbiol* 142: 23–27

45 Matsumoto H, Tazaki T, Hosogaya S (1973) A generalized transducing phage of *Serratia marcescens*. *Jpn J Microbiol* 17: 473–479

46 Humphrey SB, Stanton TB, Jensen NS, et al. (1997) Purification and characterization of VSH-1, a generalized transducing bacteriophage of *Serpulina hyodysenteriae*. *J Bacteriol* 179: 323–329

47 Schroeder CJ, Pattee PA (1984) Transduction analysis of transposon Tn551 insertions in the trp-thy region of the *Staphylococcus aureus* chromosome. *J Bacteriol* 157: 533–537

48 Burke J, Schneider D, Westpheling J (2001) Generalized transduction in *Streptomyces coelicolor*. *Proc Natl Acad Sci U S A* 98: 6289–6294

49 Stuttard C (1983) Localized hydroxylamine mutagenesis, and cotransduction of threonine and lysine genes, in *Streptomyces venezuelae*. *J Bacteriol* 155: 1219–1223

50 Stuttard C, Atkinson L, Vats S (1987) Genome structure in *Streptomyces* spp.: adjacent genes on the *S. coelicolor* A3(2) linkage map have cotransducible analogs in *S. venezuelae*. *J Bacteriol* 169: 3814–3816

51 Suss F, Klaus S (1981) Transduction in *Streptomyces hygroscopicus* mediated by the temperate bacteriophage SH10. *Mol Gen Genet* 181: 552–555

52 Hava DL, Camilli A (2001) Isolation and characterization of a temperature-sensitive generalized transducing bacteriophage for *Vibrio cholerae*. *J Microbiol Methods* 46: 217–225

53 Ichige A, Matsutani S, Oishi K, et al. (1989) Establishment of gene transfer systems for and construction of the genetic map of a marine *Vibrio* strain. *J Bacteriol* 171: 1825–1834

54 Muramatsu K, Matsumoto H (1991) Two generalized transducing phages in *Vibrio parahaemolyticus* and *Vibrio alginolyticus*. *Microbiol Immunol* 35: 1073–1084

55 Weiss BD, Capage MA, Kessel M, et al. (1994) Isolation and characterization of a generalized transducing phage for *Xanthomonas campestris* pv. campestris. *J Bacteriol* 176: 3354–3359

56 Meile L, Abendschein P, Leisinger T (1990) Transduction in the archaebacterium *Methanobacterium thermoautotrophicum* Marburg. *J Bacteriol* 172: 3507–3508

7 Use of Conditional-replication, Integration, and Modular CRIM Plasmids to Make Single-copy *lacZ* Fusions

Lu Zhou, Soo-Ki Kim, Larisa Avramova, Kirill A. Datsenko and
Barry L. Wanner

Contents

1 Introduction

Ever since a fusion of the *lac* operon to a foreign promoter (for a purine biosynthetic gene) was shown to subject the *lacY* gene to exogenous control (by purines [1]), *lac* reporter fusions have been proven to be invaluable to innumerable studies of gene structure, gene regulation, protein localization,

Methods and Tools in Biosciences and Medicine
Prokaryotic Genomics, ed. by M. Blot

protein folding, protein-protein interactions, and other aspects of cell biology [2]. Consequently, many protocols and vectors have been developed over the years for the construction of *lac* fusions, not only in *E. coli* and other bacteria but also in many other cell types [3–5].

The first *lac* fusion resulted from a genetic selection in which the *lac* operon had apparently become joined to an unregulated (constitutive) promoter that happened to lie upstream [6]. The *purE-lacY* fusion cited above was isolated afterwards by using a merodiploid strain with *purE* and the *lac* operon on an F' factor. This arrangement permitted creating a fusion of more distantly located genes without deletion of an essential gene from the chromosome. Several ingenious methods have subsequently been devised to transpose the *lac* operon, the target gene, or both to nearby one another to facilitate formation of the respective fusions [7, 8]. The development of the Mu-*lac* transposon allowed creating *lac* operon fusions in one step [9], thus extending *lac* fusion technology to many other genes in *E. coli* as well as other bacteria. Since these fusions were generated on an F' factor or the *E. coli* chromosome, they existed in single-copy and were especially useful in gene regulation studies.

With the advent of recombinant DNA technology, many plasmid vectors were made for generating fusions of *lacZ* and other reporter genes to sequences from any organism, essentially at will. Most vectors for generating *lac* reporter fusions in bacteria that have gained popularity are multicopy plasmids [10–12]. The pRS and pTL series have been used most extensively because the resultant fusions can easily be recombined into the *E. coli* chromosome in single-copy. This is accomplished in two steps. First, the fusion is transferred by homologous recombination onto a specialized λ phage, e. g., λRS45 [11] or λRZ5 (unpublished phage from R. Zagursky described in [13]), which has two properly aligned regions of homology in common with these vectors. Second, a host is lysogenized with the recombinant phage and lysogens are screened for ones with a single copy of the phage. Even though additional steps are required to isolate the recombinant phages and to make and screen the lysogens, strains with fusions on the chromosome are preferable in gene regulation and physiology studies to avoid copy-number artifacts.

There are also now many ways for making *lacZ* fusions by using transposons. Transposition is conveniently done *in vivo* by delivering the transposon on a "crippled" suicide phage or conjugative plasmid that cannot replicate in the infected or recipient host (e. g., [14–18]). In addition, several highly efficient *in vitro* transposition systems have been developed recently [19–22]. These systems essentially permit the isolation of mutants or fusions simply by transforming cells with a reaction mixture containing the transposase, an appropriately designed donor (transposon) DNA molecule, and the target DNA. Using transposons to isolate fusions has the advantage that the fusions usually reside in single-copy on the chromosome. Therefore, the target gene is interrupted for which a phenotype can aid in its identification. Nowadays, the target gene and the precise insertion site usually can be readily identified by a variety of standard molecular techniques.

In this chapter, we describe how we have used conditional-replication, integration, and modular (CRIM) plasmids [23] to construct *lacZ* reporter fusions. These reporter CRIM plasmids have all the advantages of traditional *lacZ* cloning vectors mentioned above, as well as other advantages. The CRIM plasmids have the γ replication origin of R6K that requires the *trans*-acting Π protein for replication. Therefore, they are propagated in special hosts that synthesize Π from a chromosomal *pir* gene. CRIM plasmids also have a phage attachment (*attP*) site, which allows them to be integrated into (or retrieved from) a bacterial attachment (*attB*) site of normal (non-*pir*) hosts at ease.

Integration and retrieval occur by site-specific recombination by supplying the phage integrase (Int) protein with or without excisionase (Xis) in *trans* from CRIM helper plasmids. Accordingly, CRIM plasmids have been useful not only for constructing specific *lacZ* fusions, which are integrated for studying in single-copy, but also for making fusion libraries, which are integrated *en masse* for screening in single-copy, after which selected ones are later easily retrieved. Since integration and retrieval occur by site-specific recombination, the original and retrieved plasmids are identical. Retrieved CRIM plasmids are then propagated as plasmids for subsequent molecular analysis or are integrated directly into the chromosomes of other hosts, e. g., to test for genetic regulation, without further *in vitro* manipulations.

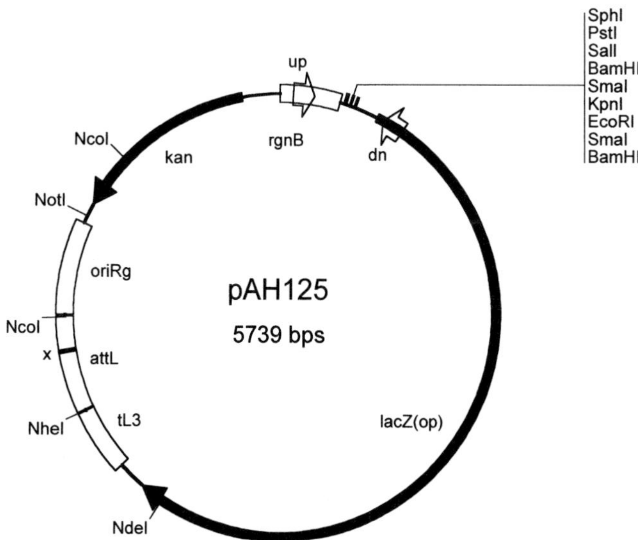

Figure 1 Basic CRIM plasmid for construction of *lacZ* (op) transcriptional fusions.

The *lacZ*, *attP*, *oriR*γ, and antibiotic resistance marker are modular to permit easy construction of new variants with different combinations of features. Because of the manner in which it was constructed, the native *Eco*R1 site of *lacZ* is absent in pAH125 [23]. An x marks the *att*λ crossover site. Sites of sequencing primers are marked as up (upstream; TTGTCGGTGAACGCTCTC-CT) and dn (downstream; AAGTTGGGTAACGCCAGG). The same primers are used to sequence promoter inserts of all *lacZ* CRIM plasmids described here: oriRg, *oriR*γ; attL, *att*λ.

The CRIM plasmids are modular [23]. Individual ones encode different resistance markers and have different *attP* sites. Therefore, multiple CRIM plasmids, e. g., one having a *lacZ* fusion and another encoding a regulator protein, can be used together in the same cell. We have most often used CRIM plasmids with an *att*λ site for making *lacZ* fusions. A typical CRIM plasmid for making *lacZ* transcriptional (also called operon [op]) fusions is shown in Figure 1. CRIM Plasmids for making *lacZ* translational (also called gene or protein [pr]) fusions are shown in Figure 2. Other CRIM plasmids for making *lacZ* (op) fusions include ones with specially engineered "zero-background" cloning sites [24] for the preparation of random fusion libraries (Fig. 3). We also describe CRIM plasmids with isopropyl β-D-galactopyranoside (IPTG) or L-arabinose-inducible promoter-*lacZ* fusions. These have been useful for conditionally expressing foreign genes at different levels.

A. Translational fusion CRIM plasmids

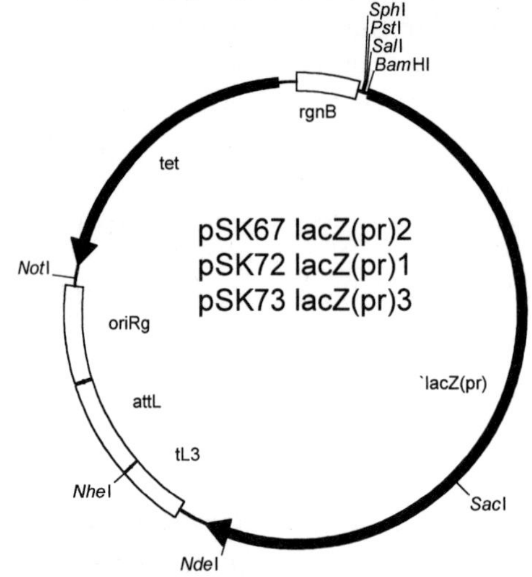

B. Fusion junction sequences

```
pSK67   G CAT GCC TGC AGG TCG ACT CTA GAG GAT CCC GTC
          h   a   c   r   s   t   l   e   d   p   V

pSK72   G CAT GCC TGC AGG TCG ACT CTA GAG GAT CAG ATC
          h   a   c   r   s   t   l   e   d   q   i
        TCC GGA TCC GAT CCC GTC
          s   g   s   d   p   V

pSK73   G CAT GCC TGC AGG TCG ACT CTA GAG GAT
          h   a   c   r   s   t   l   e   d
        CGG ATC CGG AGA TCT GAT CCC GTC
          r   i   r   r   s   d   p   V
```

Figure 2 CRIM plasmids for construction of *lacZ* (pr) translational fusions. Panel A shows the structure of pSK67, which is identical to pSK72 and pSK73 except for the polylinker preceding `*lacZ* (pr). In-frame translational fusions can be generated by using any combination of polylinker sites. pSK72 and pSK73, but not pSK67, also have a *Bgl*II site in the polylinker. Genbank accession number of pSK67 is in the Materials section. Panel B shows the sequence from the *Sph*I site to residue 13 (V) of normal, full-length LacZ. Lowercase residues are encoded by the fusion junction. The *Sph*I and *Bam*HI sites are underlined.

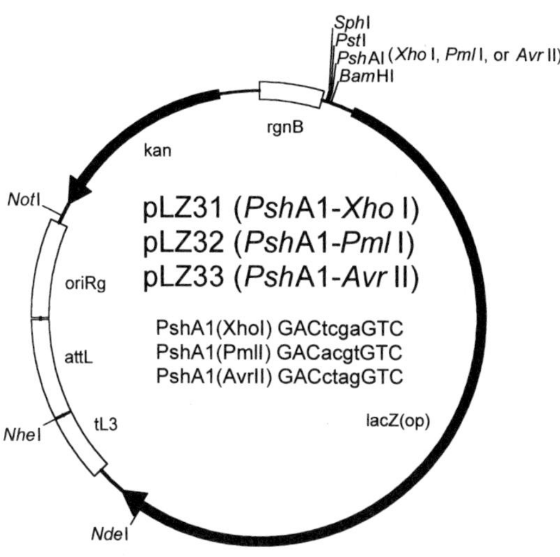

Figure 3 Zero-background CRIM plasmids for construction of *lacZ* (op) transcriptional fusions. pLZ31, pLZ32, and pLZ33 are derivatives of pAH125 (Fig. 1), into which variant *PshA1* (*Xho*I), *PshA*I (*Pml*I), and *PshA*I (*Avr*II) sites were introduced. The sequences between the *Pst*I and *Bam*HI sites are shown.

2 Materials

2.1 Bacteria and plasmids

Strains are described in Table 1. Our basic *lacZ* reporter CRIM plasmid is shown in Figure 1. BW25141 (*pir+*) and BW25142 (*pir-116*) are used to propagate CRIM plasmids at medium (15 per cell) or high (250 per cell) copy number, respectively. CRIM helper plasmids for integration into and retrieval from *att*λ are shown in Figure 4. All CRIM helper plasmids are listed in Table 2. Other *lacZ* reporter CRIM plasmids are shown in figures as cited in the text.

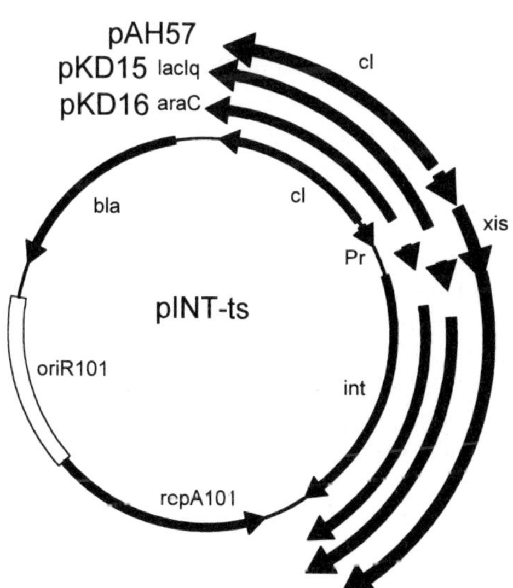

Figure 4 CRIM helper plasmids. Int and Xis/Int helper plasmids for *att*λ are shown. A complete list of CRIM helper plasmids is in Table 2. All CRIM helper plasmids are low-copy number plasmids that are temperature-sensitive for replication and hence easily curable. Most are similar to pINT-ts (or pAH57) and express *int* alone (or with *xis*) from λ*p*r under *cI857* control, which is also borne by the plasmids. pKD15 expresses *int* behind UV5i under LacI control, which is encoded by pKD15, while pKD16 expresses *int* behind *araBp* under AraC control, which is encoded by pKD16. Only relevant regions are shown. UV5i is a variant of *lacUV5* with an idealized upstream operator.

Table 1   Bacterial strains

Name[a]	Description[b]	Line[c]	Source
BW22653	*lacI*q *rrnB3* Δ*lacZ4787 rph-1*	BD792	[25]
BW22831	*lacI*q Φ(Δ*lacP rrnB3 araBp-lacZ*(op))*1542 rpo-S396*(Am) DE(*araBAD*)*567 rph-1*	BD792	[26]
BW25113	*lacI*q *rrnB3* Δ*lacZ4787 hsdR514* DE(*araBAD*)*567* DE(*rhaBAD*)*568 rph-1*	BD792	[23, 25, 27, 28]
BW25141	*lacI*q *rrnB3* Δ*lacZ4787* Δ*phoBR580 hsdR514* DE(*araBAD*)*567* DE(*rhaBAD*)*568 galU95* Δ*endA9* Δ*uidA3::pir*+ *rph-1 recA1*	BD792	[23, 25, 27, 28]
BW25142	*lacI*q *rrnB3* Δ*lacZ4787* Δ*phoBR580 hsdR514* DE(*araBAD*)*567* DE(*rhaBAD*)*568 galU95* Δ*endA9* Δ*uidA4::pir-116 rph-1 recA1*	BD792	[23]
BW25993	*lacI*q *hsdR514* DE(*araBAD*)*567* DE(*rhaBAD*)*568 rph-1*	BD792	[27]
BW28357	*lacI*q *rrnB3* Δ*lacZ4787 hsdR514* DE(*araBAD*)*567* DE(*rhaBAD*)*568 rph*+	BD792	This laboratory, unpublished
BW29716	*lacI*q Δ*lacZ1229::cat lacP-lacY hsdR514* DE(*araBAD*)*567* DE(*rhaBAD*)*568 rph*+	BD792	This laboratory, unpublished
BW29717	*lacI*q Δ*lacZ1230 lacP-lacY hsdR514* DE(*araBAD*)*567* DE(*rhaBAD*)*568 rph*+	BD792	Cm[s] of BW29716 with pCP20
BW29718	*lacI*q Δ*lacZ1229::cat lacP-lacY rph*+	MG1655	This laboratory, unpublished
BW29720	*lacI*q Δ*lacZ1230 lacP-lacY rph*+	MG1655	Cm[s] of BW29718 with pCP20
BW29721	*lacI*q *rrnB3* Δ*lacZ4787 rph*+	MG1655	This laboratory, unpublished

[a]   All strains are λ⁻ and F⁻ derivatives of *E. coli* K-12.

[b]   All known mutations are indicated. The *rph-1* and *rph*+ alleles refer to a frameshift mutation [29] and the corrected allele, respectively. Strains are denoted as *rph*+ if they are descended from ones in which this mutation has been corrected. A mutant carrying the Δ*lacZ1229::cat lacP-lacY* fusion (SC824, from S. Chiang) was made by electroporation of BW25993 carrying pKD46 with a PCR product generated with pKD3 as template essentially, as described elsewhere [27], by using PCR primers designed to remove *lacZ* entirely and to express *lacY* behind *lacP* under LacI control (S. Chiang and E. Rubin, personal communication). This mutation was then transferred by P1*kc* transduction to make BW29716 and BW29718, after which the FRT-flanked *cat* cassette was evicted by using pCP20 to generate the Δ*lacZ1230 lacP-lacY* fusion.

[c]   The line corresponds to the strain name when received from another lab. BD792 (CGSC6159) and MG1655 (CGSC6300) are in turn F- derivatives of W1485F+ that were isolated by Bruce Duncan and Mark Guyer, respectively (B. Bachman, personal communication). We showed that both contain the *rph-1* frameshift mutation that exists in many lines of wild-type *E. coli* K-12 [29], which we corrected for those identified as being *rph*+ (K. A. Datsenko and B. L. Wanner, unpublished data). BD792 has the *rpoS396* (Am) mutation at codon for Q33 of RpoS (K. A. Datsenko and B. L. Wanner, unpublished data), which has been corrected in ancestors of all BD792 derivatives listed, except BW22831.

Table 2   CRIM helper plasmids[a]

attP site	Int helper plasmid[b]	Xis/Int helper plasmid[c]
*att*λ	pINT-ts, pKD15, pKD16	pAH57
*att*Φ80	pAH123	pAH129
*att*HK022	pAH69	pAH83
*att*P21	pAH121	pAH122
*att*P22	pAH130	pAH131

[a]   Genbank accession numbers are in the Materials section.
[b]   pAH69, pAH121, pAH123, and pAH130 are similar to pINT-ts (Fig. 4), except that each encodes the respective Int protein.
[c]   pAH83, pAH122, pAH129, and pAH131 are similar to pAH57 (Fig. 4), except that each encodes the respective Xis and Int proteins.

2.2   Nucleotide sequence accession numbers

GenBank accession numbers for the CRIM plasmids are: AY054372 (pAH125), AY054373 (pLA1), AY150261 (pLA2), AY150262 (pLA4), AY150263 (pLA5), AY150264 (pLA7), AY150265 (pLA8), AY150266 (pLA9), AY150267 (pLZ31), and AY150268 (pSK67). GenBank accession numbers for the CRIM helper plasmids are: AY048715 (pAH57), AY048718 (pAH69), AY048720 (pAH83), AY048724 (pAH121), AY048725 (pAH122), AY048726 (pAH123), AY048727 (pAH129), AY048728 (pAH130), AY048729 (pAH131), AY048741 (pINT-ts), AY150269 (pKD15), and AY150270 (pKD16). GenBank accession number for the Δ*lacZ1230 lacP-lacY* fusion junction is AF548059.

2.3   Chemicals, media, buffers, and other supplies

5-Bromo-4-chloro-3-indolyl-β-D-galactopyranoside (X-gal) is from Bachem (Torrance, CA); 1-S-octyl-β-D-thioglucoside (OTG) is from Anatrace (Maumee, OH); and *o*-nitrophenyl β-D-galactopyranoside (oNPG), IPTG, L-arabinose, antibiotics, and ethylene glycol *bis*(β-aminoethyl ether) N,N,N',N'-tetraacetic acid (EGTA, free acid) are from Sigma (St. Louis, MO). X-Gal is dissolved in dimethylformamide (DMF) at 40 mg/ml. When stored refrigerated in the dark, it is stable for about one month. A 1 *M* OTG stock is stored frozen (–20 °C), thawed, and held on ice until use. A stock solution of 4% oNPG in water is stored frozen (–20 °C) indefinitely and used to make other oNPG solutions for assay. Other chemicals are from routine sources.

Luria-Bertani (LB) broth (without glucose), tryptone-yeast extract (TYE) agar (pH 7.0), M63, and MOPS media are prepared as described elsewhere [30]. Media are solidified using Bacto agar (1.5% final). Nutrient broth (for freezing vials) contains 10 g Bacto Nutrient broth and 8 g NaCl per L. SOB [31] medium contains 2% tryptone, 0.5% yeast extract, 10 mM NaCl, 2.5 m*M* KCl, 10 m*M*

$MgCl_2$, and 10 mM $MgSO_4$. It is prepared without Mg^{2+} and autoclaved. This is done by dissolving 20 g tryptone (Bacto), 5 g yeast extract (Bacto), and 0.5 g NaCl in ca. 950 ml water; adding 10 ml 0.25 M KCl, adjusting the pH to 7.0 with 1 N NaOH; adjusting the volume to 1 L; and autoclaving. A 2 M Mg^{2+} stock solution (1 M $MgCl_2 \cdot 6H_2O$ and 1 M $MgSO_4 \cdot 7H_2O$) is prepared and filter sterilized. To prepare SOB, 100 µl of the 2 M Mg^{2+} stock solution is added per 10 ml of the Mg^{2+}-free SOB immediately before use. SOC medium is SOB with 20 mM D-glucose. SOC is made by adding 200 µl of filter-sterilized 1 M glucose per 10 ml SOB.

X-Gal is the blue dye for detection of β-galactosidase. Agar plates containing X-Gal are unstable and should be used within one week. When many X-Gal plates are prepared, the molten agar after autoclaving is cooled to 55°C, other ingredients are added, as necessary, and then 1 ml of the X-gal stock solution is added per liter. When only a few X-Gal plates are required, 75 µl of the X-Gal stock solution is spread onto the surface with a glass or metal spreader and a turntable or by using several (ca. 20 to 30) sterile, 3 mm soda lime glass beads and manually tilting the plate.

When following the growth of many cultures simultaneously, samples for cell density measurements are removed periodically and added to "fix" (0.5 ml formaldehyde per 100 ml water) to prevent further growth. When using a spectrophotometer, 0.5 ml samples are added to tubes containing 1.0 ml fix, the tubes are mixed by vortexing, and the diluted samples are later read. When using a microplate reader, 100 µl portions are pipetted directly into wells of a 96-well microplate containing 150 µl fix. Samples are mixed by pipetting and left at room temperature until they are read. Samples with an A_{410} > ca. 0.7 are further diluted to obtain accurate readings. All wells should contain the same total volume (250 µl) when readings are taken.

Z buffers for microtiter plate assays include: (1) Z stock buffer (pH 7.5), which contains 11.6 g anhydrous Na_2HPO_4, 2.48 g $NaH_2PO_4 \cdot H_2O$, 0.75 g KCl, and 0.246 g $MgSO_4 \cdot 7H_2O$ per liter and is stored refrigerated; (2) Z assay buffer, which additionally contains 1% β-mercaptoethanol (β-ME), 100 µg/ml chloramphenicol (Cm), and 15 mM OTG; and (3) Z holding buffer, which contains in addition 10% glycerol.

Microelectroporator chambers (for a BRL Cell-Porator) are from Labrepco (Horsham, PA). Standard 96-well flat-bottom polystyrene microtiter plates with covers (Evergreen Scientific 290–8115–01F) and microplate sealing tape are from Life Science Products (Denver, CO).

2.4 Equipment

A variety of microplate readers, electroporators, and PCR machines are suitable. Protocols described in this chapter were carried out using a BRL Cell-Porator with Voltage Booster, Molecular Devices 340PC plate reader, and an MJ

Research PTC-200 Cycler. Temperatures, times, and volumes may require adjustments with equipment of different models or manufacturers.

3   Methods

Protocol 1   Bacteria preservation

1. Bacteria are routinely stored at –70 °C in nutrient broth (preferred) or LB broth containing 8% dimethyl sulfoxide (DMSO).
2. A one-half dram (2 ml) glass vial containing 1 ml nutrient broth (to which an antibiotic is freshly added for plasmid-bearing strains) is inoculated from a fresh isolated colony of a new strain.
3. The vial is incubated at 30 °C on a roller until saturation (ca. 16 h).
4. Then 90 µl DMSO (containing 0.5 ml 95% ethanol per 100 ml DMSO) is added, and the vial is mixed briefly and placed directly into the ultracold freezer.

Protocol 2   Bacteria recovery

1. Bacteria are recovered without thawing by scraping the surface with a toothpick or pipet tip and colony-purified by streaking onto an appropriate agar medium (usually TYE agar without an antibiotic, except for plasmid-bearing strains).
2. For physiology experiments, cells are adapted on agar of a similar composition prior to inoculation into liquid media.
3. Cells are routinely passaged on M63 agar before growing them on MOPS agar.
4. Typically, a strain is revived on TYE agar and isolated colonies are re-streaked on glucose M63 agar and then on glucose MOPS agar before inoculation of glucose MOPS broth.
5. Colonies from the glucose MOPS agar are streaked onto MOPS agar with different carbon sources prior to inoculation into the respective media.
6. For inoculation, an isolated colony is suspended in ca. 50 µl of saline (0.85% NaCl) and ca. 10 µl portions are inoculated into culture tubes (or flasks) containing similar media without or with inducer.
7. All strains are assayed in triplicate by suspending three colonies separately and inoculating one tube (or flask) with and without inducer with each suspension.
8. 1 ml cultures are grown in 18 mm culture tubes with closures to permit air exchange in a tube roller.
9. Larger volumes are grown in shake flasks.

Protocol 3 Preparation of electrocompetent cells

1. To make electrocompetent cells, 1.5 ml culture is transferred to a microfuge tube and placed on ice for 2 min and the cells are collected by centrifugation for 1 min (at maximum speed in a high-speed microfuge at 4 °C).
2. The supernatant is discarded and the cells are washed three times with 1 ml ice-cold 10% glycerol by resuspension and 15 to 30 s centrifugations.
3. Care is taken to remove the supernatants promptly because the cell pellets are soft.
4. After three washings, the cells are resuspended with 50 µl ice-cold 10% glycerol and held on ice until use.
5. This yields a total of ca. 80 µl, or enough for four electroporations.
6. The protocol is scaled up to prepare larger amounts of electrocompetent cells by using larger tubes and a high capacity centrifuge.
7. Cells prepared in this way are routinely stored in 20 or 40 µl aliquots at –70 °C.
8. Frozen electrocompetent cells are thawed in an ice-water bath immediately before use.

Protocol 4 Electroporation

1. For electroporation, 20 µl electrocompetent cells and 1 µl DNA (50 ng) are combined and placed into the electroporation cuvette.
2. The Cell-Porator has a cuvette holding chamber that is filled with an ice-water mixture so the cells are held at 0 °C throughout the process.
3. Electroporation is done as recommended by the manufacturer.

Protocol 5 CRIM plasmid integration

Integration can be done by introducing the CRIM plasmid into electrocompetent or chemically competent cells [31]. We nearly always use electroporation solely because it is more efficient. Our protocol may require minor adjustments when using chemical transformation.

1. Cells carrying the respective CRIM helper plasmid are prepared in advance by selecting ampicillin-resistant transformants at 30 °C.
2. For preparation of electrocompetent cells, a fresh isolated colony is inoculated into a small flask containing 5 ml SOB (without Mg^{2+} and with ampicillin [100 µg/ml]) and incubated in a 30 °C shaking water bath.
3. For integration into $att\lambda$ pINT-ts transformants are grown to an A_{600} of ~0.5–0.6 (ca. 5 to 6 h) and the flask is then shifted to a 42 °C shaking water bath for ca. 20 min longer and then placed on ice. Growth in water baths is preferred to permit rapid temperature shifts.
4. Alternatively, pKD15 and pKD16 transformants are grown to an A_{600} of ~0.2–3.0 (ca. 4 to 5 h), then 1 mM IPTG (for pKD15) or 1 mM L-arabinose (for pKD16) is added, incubation is continued ca. 1 to 1.5 h longer to an A_{600} of ~0.6–0.7, and the flasks are placed on ice.

5. Cells are then made electrocompetent as described above.
6. Following electroporation, the cell-DNA mixture is added to 1 ml SOC (no antibiotic) in an 18 mm tube, incubated at 42 °C for 30 min, then at 37 °C for 1 to 1 1/2 h longer.
7. Portions are spread onto media that are selective for the CRIM plasmid and lack ampicillin to permit loss of the helper plasmid.
8. Single-copy integrants of the *lacZ* CRIM plasmids described here are usually selected on TYE agar containing X-gal and kanamycin at 10 µg per ml or tetracycline at 8 µg per ml.
9. Concentrations used with CRIM plasmids encoding different resistances are given elsewhere [23].
10. After spreading, the plates are incubated 16 to 20 h at 37 °C. Cells are colony-purified without antibiotic selection before testing for loss of the helper plasmid (by testing for ampicillin sensitivity).

Protocol 6   PCR test of integrant copy number

1. Isolated colonies are picked up with a plastic tip or glass capillary (but not with a toothpick!) and suspended in 20 µl water.
2. Combine 5 µl of the cell suspension, 10 pmol of each primer (P1 to P4; Fig. 4), and 0.5 U of Taq DNA polymerase (New England Biolabs) in 1 X PCR buffer-2.5 mM MgCl$_2$ with deoxynucleoside triphosphates in a final volume of 20 µl.
3. PCR is carried out for 25 cycles (denaturing for 1 min at 94 °C, annealing for 1 min at 63 °C [for integration at *att*λ], and extending for 1 min at 72 °C).
4. The same P2 and P3 primers are used with all CRIM plasmids, while the P1 and P4 primers are specific for the chromosomal *attB* site (Fig. 5).
5. Single PCR reactions are run with all four primers in the same tube.
6. Different annealing temperatures and P1 and P4 primers are used for CRIM plasmids that integrate at different *attB* sites, as described elsewhere [23].
7. Usually three or four colonies are picked directly from the selection plate and initially tested by PCR for single-copy ones.
8. Two or three single-copy candidates are then colony-purified nonselectively once or twice and retested by PCR to be sure that they are pure and stable.

Protocol 7   CRIM plasmid retrieval

1. Retrieval is carried out by transduction or transformation of a recipient that is *pir*+, for replication of the CRIM plasmids, *recA*, to avoid homologous recombination events, and that carries the respective Xis/Int CRIM helper plasmid. Accordingly, the CRIM plasmid enters the recipient as a linear DNA molecule attached to chromosomal DNA of the donor. Upon entry into the recipient, Xis and Int catalyze the excision and recircularization of the CRIM plasmid, which then replicates as a free plasmid. In essence, site-specific

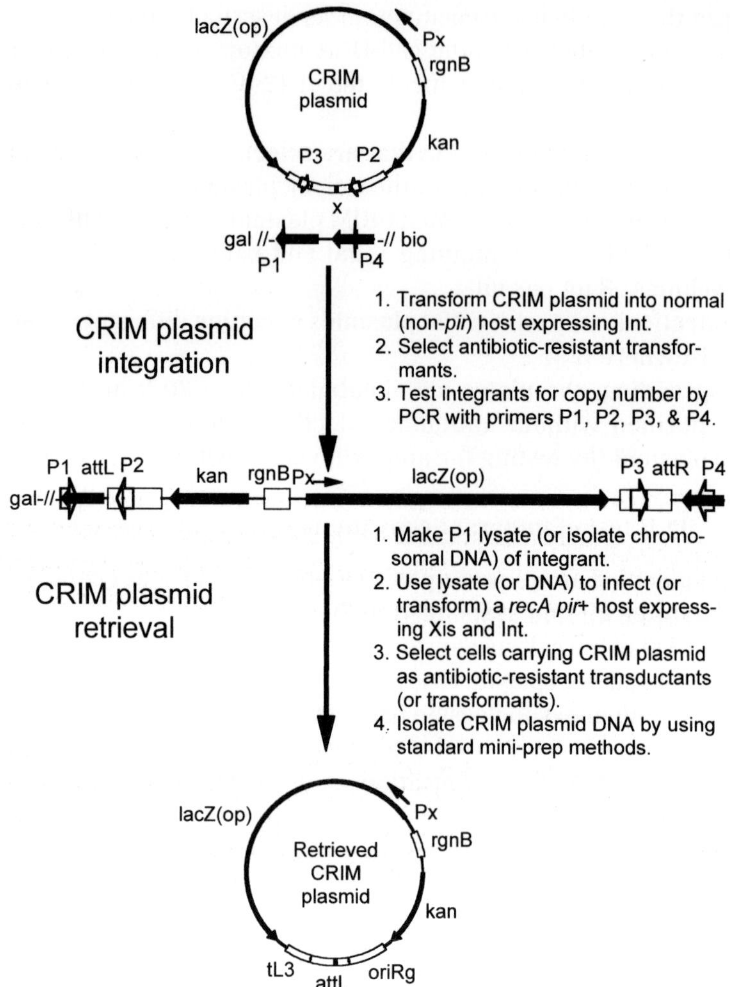

Figure 5   Integration and retrieval of CRIM plasmid. CRIM plasmids are integrated into a normal (non-*pir*) host under conditions of Int synthesis. Integration of an *attλ* CRIM plasmid occurs by site-specific recombination at *attB*, which lies between the chromosomal *gal* and *bio* loci. Integrates are selectable as antibiotic-resistant transformants because the CRIM plasmids cannot replicate in the absence of Π. When using the P1 to P4 primers, a control strain with an empty *attB* site yields a single *attB* (741-nt, P1/P4) PCR product, and those with a single integrated CRIM plasmid yield both *attL* (577-nt, P1/P2) and *attR* (666-nt, P3/P4) PCR products. Ones with multiple CRIM plasmids integrated in tandem yield three products: (1) an *attL* (577-nt, P1/P2) PCR product, (2) an *attR* (666-nt, P3/P4) PCR product, and (3) a CRIM plasmid-specific (502-nt, P2/P3) PCR product. (An example of a multiple integrant is shown in [23]). Integration also can occur by homologous recombination (see Troubleshooting section), although this is infrequent. These recombinants are recognizable as ones with an *attλ* (741-nt, P1/P4) and a CRIM plasmid-specific (502-nt, P2/P3) PCR products. The latter is seen because the *attλ* region of the CRIM plasmid is uninterrupted in such recombinants. Primer sequences are: P1, GGCATCACGGCAATATAC; P2, ACTTAACGGCTGACATGG; P3, ACGAGTATCGAGATGGCA; and P4, TCTGGTCTGGTAG-CAATG.

recombination results in cloning the CRIM plasmid from the chromosome. Because we usually do this by using the generalized transducing phage P1*kc*, we have termed the process PIX cloning for P1, Int, Xis cloning [32].

2. P1*kc* lysates are made on the integrants by standard procedures [30].

3. To make lysates, recipient cells are grown in 1 ml LB with ampicillin at 30 °C to early stationary phase in 18 mm tubes (in a roller).

4. Cells are collected by centrifugation and resuspended in 10 mM MgCl$_2$, 2.5 mM CaCl$_2$.

5. A portion of the lysate (usually 5 µl) is added to 100 µl cell suspension, and the mixture is held at room temperature.

6. After 20 min, 1 ml LB containing 10 mM EGTA (neutralized with NaOH) is added, the tube is vortexed, and the cells are collected by centrifugation.

7. Cells are resuspended in 1 ml LB with 10 mM EGTA (no antibiotic), incubated at 37 °C for 1 h, at 42 °C for 30 min, and at 37 °C for an additional hour.

8. Portions are then spread onto TYE agar with an antibiotic to select for the CRIM plasmid (without ampicillin) and incubated at 37 °C. The protocol and strains used for the recovery of CRIM plasmids by DNA transformation are described elsewhere [23].

Protocol 8   Simple plate test for estimation of β-galactosidase activity

X-Gal is a very sensitive indicator of β-galactosidase activity, however it is not very quantitative. To estimate relative β-galactosidase levels among different strains, we use oNPG instead. In the past, we did this by dripping a solution of 0.4% *o*NPG in Z buffer containing 0.05% SDS (without β-ME or Cm) onto colonies grown on an agar medium [33]. Since OTG is a more effective lysing agent than SDS under these conditions, we now routinely do this by using instead a solution of 0.4% *o*NPG in Z buffer containing 15 mM OTG. For comparisons, appropriate negative (Δ*lac*) and positive (Lac$^+$ constitutive) control strains are tested side by side. The 0.4% *o*NPG Z buffer solution is stored refrigerated for up to one week.

Protocol 9   Cell growth and microplate β-galactosidase assays

Our standard protocols for measuring cell growth and β-galactosidase activities when using test tubes and a spectrophotometer are described in detail elsewhere [30]. A microplate reader facilitates assaying many more samples and the use of smaller volumes. While standard spectrophotometric and microplate β-galactosidase assays are basically similar, a few differences are notable. For tube assays, cells are usually lysed by treatment with SDS and chloroform. We use instead 15 mM OTG to lyse cells for microplate assays because it is rapid and more effective than other lysis methods. Tube assays are usually stopped before reading by adding ca. one-half volume of 1 M Na$_2$CO$_3$, which raises the pH to ca. 10. Since tubes with more enzyme turn yellow more rapidly, they are stopped earlier and the times are noted. For simplicity, microplates are instead read continuously without stopping the reactions until all samples turn sufficiently yellow. Although all data are collected automatically, we use only those absorbance readings within a set range (ca. 0.05 to 0.7) to calculate the results. It should be noted that the absorbance of o-nitrophenol changes with pH. There is a ca. 2.5-fold increase from pH of 6.9 to 8.6, while β-galactosidase activity is only modestly affected between pH 6.9 and 7.6 (data not shown). Therefore, to enhance color development of o-nitrophenol, we use Z buffers at pH 7.5 (instead of pH 7.0) for microplate β-galactosidase assays. It is also important that all wells contain the same total volume. We run assays with a total volume of 250 µl per well. Cross-contamination also can be a serious problem if assays are run for long periods (> ca. 2 h). For short periods, plates are covered with standard lids between readings. For samples requiring long incubations, the wells are sealed with microplate sealing tape in order to prevent cross-contamination resulting from vaporization and condensation of o-nitrophenol. Our detailed protocol follows:

1. Cultures are sampled by removing 100 µl portions. For absorbance readings, 100 µl samples are added directly to 150 µl fix in wells of a microplate and mixed by pipetting. For enzyme assay, a second 100 µl sample is added to a microfuge tube containing 200 µl Z assay buffer and the tube is vortexed. Accordingly, the dilution factors for the cell density measurements and β-galactosidase assays are 2.5 and 3.0, respectively (step 9). The microplate is left at room temperature. The enzyme samples are kept in a floating microfuge rack in an ice-water bath (0 °C) until all data are collected and evaluated (step 7). If these samples are to be kept more than several hours before they are assayed, Z holding buffer is used instead.

2. The absorbance values of the "fixed samples" are measured in a plate reader at A_{410}. If values exceed ca. 0.7, further dilutions are made as necessary. The dilution factor is corrected accordingly. For convenience, the cell densities and β-galactosidase assays (o-nitrophenol production) are both read at A_{410}.

3. To measure β-galactosidase, 30 µl portions of the enzyme samples are added to microplate wells containing 200 µl Z assay buffer. Extreme care should be taken to avoid air bubbles, which can affect the readings. Once substrate is added (step 6), the total assay volume is 250 µl, so the sample dilution factor is 8.33. Depending upon the cell density and amount of enzyme, the portion assayed is varied. For cultures with an actual A_{410} between 0.2 and 1.5, 30 µl portions are usually adequate. The amounts are adjusted as necessary so that the cell absorbance in the wells is ca. 0.02 or less. Even though OTG treatment rapidly permeabilizes cells to oNPG and reduces their turbidity, the cell densities continue to decrease for many hours. For initial values < 0.02, these changes are negligible even after 16 h incubations. We also vary the portions assayed so that the assay time is at least 10 min and generally less than 16 h (step 7). If portions are varied, the amounts of Z assay buffer are changed to maintain a total assay volume after substrate addition of 250 µl.

4. The microplate samples are read to obtain an initial value of the OTG-treated cells. Because of the OTG treatment, these values are less than the values of the fixed samples, even after correcting for dilution. The actual differences vary and depend on the time of OTG treatment, culture condition, and strain. This is why we use an initial cell density of the OTG-treated cells < 0.02. If samples with higher cell densities are assayed, additional controls (OTG-treated cells without substrate) are run in parallel for each sample.

5. The microplate with the OTG-treated cells is incubated at 28 °C for 10 min. During this time, the *o*NPG substrate solution is also pre-warmed to 28 °C. With a temperature-controlled reader and a single microplate, the reader itself can be used for incubation. When assaying several microplates, they are incubated on a metal surface inside a constant-temperature incubator nearby.

6. The assay is started by adding 20 µl 0.4% *o*NPG to each well with a multichannel pipettor. When doing kinetic assays, the actual times of the first (t_0, time zero) and last (t_e, time end) readings are arbitrary.

7. Plates are read at 5, 10, and 20 min and at later times until all samples reach an A_{410} of 0.2 to 0.6 or 16 h has passed, whichever occurs sooner. If samples turn yellow too quickly, e. g., reach an $A_{410} > 0.8$ in less than 10 min, they are re-assayed by using smaller portions of the enzyme samples.

8. Blanks containing Z assay buffer with *o*NPG but no cells are always run. As mentioned in step 4, controls for OTG-treated cells (without substrate) are sometimes also necessary.

9. Enzyme units are calculated in terms of activity (nanomoles per min) per ml of culture and specific activity (activity per cell density or mg protein) using the molar extinction coefficient, e_{410}, for *o*-nitrophenol of 4500. After subtraction of the blank, results are calculated using the following equations:

$$\text{Activity (nmoles/ml/min)} = 3.0 \text{ (CDF)} \cdot [A_{410}(t_e) - A_{410}(t_0)] \cdot 8.33 \text{ (SDF)} \cdot 222 \text{ (CF)}/(t_e - t_0),$$

where CDF is the culture dilution factor (3.0 in step 1), SDF is the sample dilution factor (8.33 in step 3), and CF is the conversion factor (222 or 10^6/4500, for conversion of absorbance to nanomoles per ml). $A_{410}(t_e)$ and $A_{410}(t_0)$ are the absorbances at the end (t_e) and start (t_0), respectively. The actual values of CDF and SDF vary depending upon the dilutions when sampling the cultures and the portions assayed.

$$\text{Specific activity} = \text{Activity} / [2.5 \text{ (FDF)} \cdot A_{410} \text{ (cells)}]$$

where FDF is the fix dilution factor (2.5 in step 1) and A_{410} (cells) is the cell density in step 2. If different volumes are taken for the cell density and enzyme samples, the appropriate values are used in the calculations.

4   Troubleshooting

Problems can arise during the construction of specific *lacZ* fusion CRIM plasmids or upon integrating them into the chromosome. While it is desirable to use high-copy-number (*pir-116*) *E. coli* hosts [34] for ease of plasmid preparation, plasmids carrying strong promoters or particular genes can be deleterious in these hosts. We have on occasion found both promoter and *lacZ* mutations in CRIM plasmids isolated from such high-copy-number hosts. We therefore routinely use medium-copy-number (*pir*+) hosts to prepare CRIM plasmid DNAs when characterizing promoters of unknown strength. We use high-copy-number (*pir-116*) hosts only to prepare CRIM plasmid DNAs when examining weak or tightly controlled promoters.

Following electroporation and selection of integrants (3.3 to 3.5), the helper plasmid is usually lost because it is unstable at 37 °C and ampicillin is omitted. If it is not lost, a few colonies are colony-purified nonselectively once at 43 °C and retested. Once integrants are chosen from the initial selective media, we routinely propagate them in the absence of an antibiotic. Since CRIM plasmids integrate by site-specific recombination, the integrants are quite stable. It is also preferable not to maintain them on an antibiotic medium to prevent inadvertent selection of multiple copy integrants, which also can occur subsequently.

We routinely test about three colonies by PCR as described in Protocol 6 and Figure 5 to find single-copy integrants, which are usually the majority. The most frequent undesirable event is the formation of multicopy (tandem) integrants at the respective *attB* site. These can predominate if too high an antibiotic concentration is used in the selective medium. We have used the concentrations in Protocol 5 to isolate single-copy integrants of hundreds of *E. coli* K-12 and a few *Salmonella typhimurium* strains. Occasionally, we varied these concentrations to reduce background growth or to find single-copy integrants.

A second undesirable event can occur if the CRIM plasmid recombines elsewhere on the chromosome and not at the respective *attB* site. These are also recognizable by the standard PCR test (Fig. 5). They probably result from homologous recombination of the CRIM plasmid with the bacterial chromosome. For example, a CRIM plasmid carrying an *E. coli* promoter-*lacZ* fusion can recombine via promoter sequences in common with the chromosome. We also frequently integrate CRIM plasmids into cells containing an integrated CRIM plasmid at a different *attB* site. In such cases, homologous recombination can occur with the resident plasmid because all CRIM plasmids have sequences in common (tL3, *oriRγ*, and *rgnB*). Importantly, these events are seldom problematic because of the high efficiency of site-specific recombination. When they occur, they are attributable to inadequate Int synthesis. The simplest remedy is to repeat the integration with newly prepared electrocompetent cells. Sometimes, a new transformant carrying the CRIM helper plasmid is used. Changing the induction protocol can also help. A CRIM helper plasmid that synthesizes Int under a different control (e. g., induction by temperature-shift or by arabinose or IPTG addition; Fig. 4) can also be advantageous.

5 Applications

Reporter CRIM plasmids are useful for both studying specific regulatory regions and screening random fusion libraries to identify new promoters. Recombinants carrying specific or random promoter-*lacZ* fusions are generated by using standard molecular biological methods [35]. To construct specific fusions, promoter fragments are conveniently synthesized by PCR using primers with 5' extensions containing restriction sites for directional cloning into a transcriptional fusion CRIM plasmid such as pAH125 (Fig. 1). Translational fusions are similarly made by using a translational fusion CRIM plasmid such as pSK67 (Fig. 2).

These plasmids also can be used for the construction of random *lacZ* fusion libraries. For example, we have used equimixtures of the translational fusion CRIM plasmids pSK67, pSK72, and pSK73 (digested with *Bam*HI) to prepare libraries with *E. coli* chromosomal DNA (digested with *Sau*3 A) to search for genes controlled by the response regulator CreB. To do this, we integrated the libraries directly into the chromosome of a Δ*lacZ* Δ*creB* host carrying a tightly regulated, rhamnose-inducible *rhaBp-crcB*⁺ fusion elsewhere on the chromosome [26]. By screening ca. 6000 colonies that were blue (Lac⁺) on X-Gal agar in the presence of rhamnose, we found 13 that were less blue in its absence. We retrieved these CRIM plasmids, sequenced their inserts, and then integrated them into the same and different hosts to study their regulation. As expected, a few contained the rhamnose-regulated *rhaS* and *rhaT* promoters. Several others contained regions upstream of *yidS* and *yieI*, suggesting that they are targets of CreB. While the goal was to unmask the role of CreB by identifying its gene targets, this was not accomplished because no role can be ascribed to YidS, YieI, or other co-transcribed genes. Nevertheless, these experiments established the utility of *lacZ* reporter CRIM plasmids for the preparation and screening of random *lacZ* fusion libraries in single-copy.

One limitation of screening random fusions is the large number of integrants that should be tested. For example, to find all promoters in a ca. 5-megabase genome at a 95% confidence limit would require testing over 50,000 integrants and at 99% confidence limits, over 250,000 integrants. These values were calculated based on the assumptions that the libraries contain random ca. 500-bp fragments and that the promoter/regulatory regions are distributed randomly around the chromosome within ca. 250-bp segments (unpublished data). Further, this number would even be larger if the libraries are "contaminated" with plasmids without inserts. To overcome this problem, we recently made new *lacZ* reporter CRIM plasmids (Fig. 3), which are based on zero-background cloning vectors developed elsewhere [24].

Conditional expression is a powerful genetic tool. It was first used to show that a control region is separable from its structural gene in the analysis of a fusion that expressed LacY under purine control [1]. Shortly afterwards, conditional (adenine-dependent) expression of this fusion was used to clone the first gene (*lacZ*) into a phage vector by directed transposition [36]. This approach has been used innumerable times since, e. g., to show that the first *trp-lacZ*, *ara-lacZ*, and *phoA-lacZ* fusions were fused to the respective promoters [8, 37, 38]. Likewise, searches for *d*amage-*in*ducible (*din*), *p*hosphate-*s*tarvation-*in*ducible (*psi*), and *in vivo in*ducible (*ivi*) genes [39–41] were logical extensions of this approach. Accordingly, many methods have been developed since that exploit conditional expression to study cell biology [42, 43].

Our basic CRIM plasmids include ones with promoters controlled by IPTG, arabinose, or rhamnose [23]. To assess their regulation, representative promoters controlled by IPTG and arabinose were cloned into an appropriate *lacZ* reporter CRIM plasmid (Figs. 6 and 7). In general, the expression of these

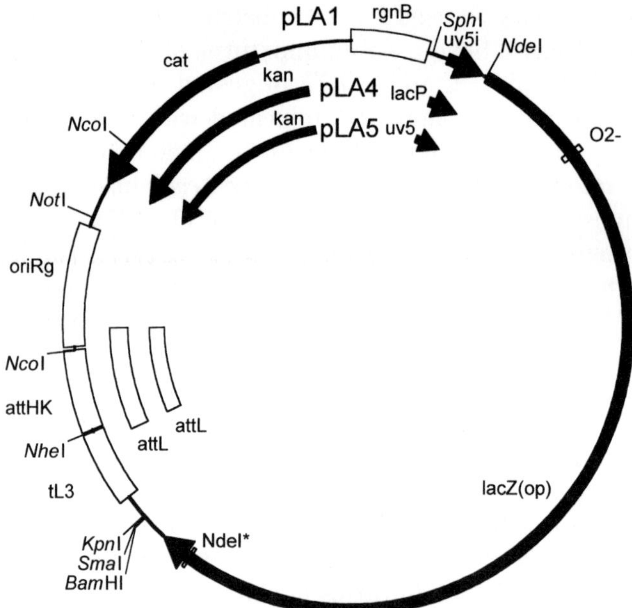

Figure 6   CRIM plasmids with IPTG-inducible promoters. Unlike the CRIM plasmids in Figures 1, 2, and 3, pLA1, pLA4, and pLA5 are derivatives of a *lacZ* CRIM plasmid in which the native *lacO2* operator was changed to the O2- sequence [44], the native *Nde*I site in *lacZ* was eliminated by a silent mutation, and a unique *Nde*I site was introduced adjacent to the Met start codon of LacZ [23]. pLA1 has an idealized upstream operator and the native *lacO1* downstream of *lac*UV5, hence the designation UV5i. In pLA4 and pLA5, *lacP* and *lac*UV5 originated as PCR fragments using as templates pUC19 [45] and pRZ6522 ([46]; from W. Reznikoff), respectively. The entire *lacZ* with the modified *lacO2* region and *Nde*I sites and all promoter segments were sequenced after initial cloning. Because of how they were constructed, the native *Eco*R1 site of *lacZ* is also absent in these plasmids. O2- and NdeI* mark the locations of the respective mutations. Genbank accession numbers are in the Materials section.

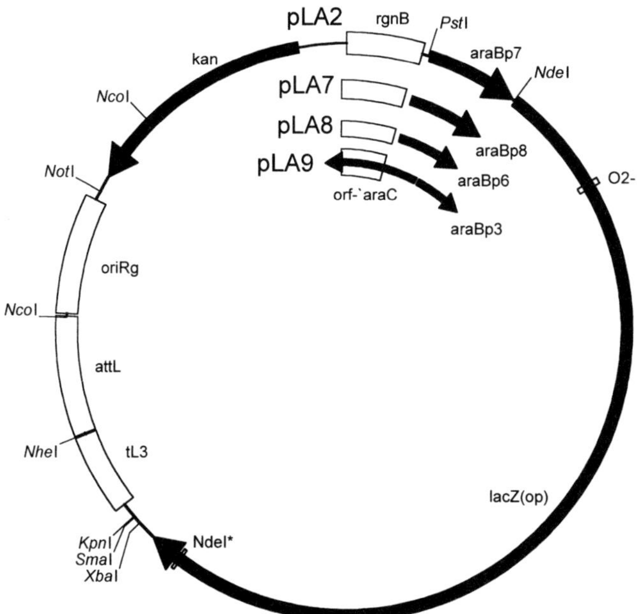

Figure 7   CRIM plasmids with arabinose-inducible promoters. pLA2, pLA7, pLA8, and pLA9 have the same *lacZ* region as those shown in Figure 6. The *araBp* segments are denoted with alleles to indicate their origin and different sequences. The *araBp3* (in pLA9) has the same region as pAH150 [23], which we have shown elsewhere to exhibit arabinose-dependent PhoB synthesis [26]. The *araBp6* (in pLA8) was generated by PCR with pAH150 as template and cloned as a 152-nt *araBp* fragment lacking native *araBpO2* [47]. The *araBp7* (in pLA2) was generated by PCR with pBAD33 [43] as template; it has the entire *araBp* regulatory region of *E. coli* B. The *araBp8* (in pLA7) was generated by PCR with strain MS1868 (from P. Youderian) as template; it has the entire *araBp* regulatory region of *S. typhimurium* LT2. As described in the text, all of these promoters, except *araBp3*, express comparable levels of β-galactosidase. An examination of the sequences revealed that pLA9 (like pAH31 and pAH150) is predicted to encode 55 N-terminal residues of AraC. Furthermore, this *araC'* segment is joined to plasmid sequences such that pLA9 and pAH150, but not pAH31, encode a 246-residue araC'-orf fusion protein resulting from translational read-through of the *rgnB* region. In contrast, pAH31 is predicted to encode an AraC' fusion protein with only a 10-residue C-terminal extension. Since the introduction of stop codons preceding *rgnB* restores high-level expression in a derivative with the same *araBp3* region (K. Zhang and B. L. Wanner, unpublished data), down-regulation of *araBp* in pLA9 (and pAH150) is a consequence of expressing this AraC'-orf fusion protein. All segments generated by PCR were sequenced after initial cloning. Genbank accession numbers are in the Materials section.

promoters is induced by IPTG or arabinose and the induced levels are similar (Fig. 8). Under these conditions, the *lac*UV5 promoter is the highest among the IPTG-regulated promoters. With one notable exception, the arabinose-regulated promoters are induced to even higher levels. The exception concerns *araBp3* in pLA9. The *araBp3-lacZ* fusion in this CRIM plasmid was unexpectedly found to express *lacZ* at only ca. 2% of the level of the identical fusion when

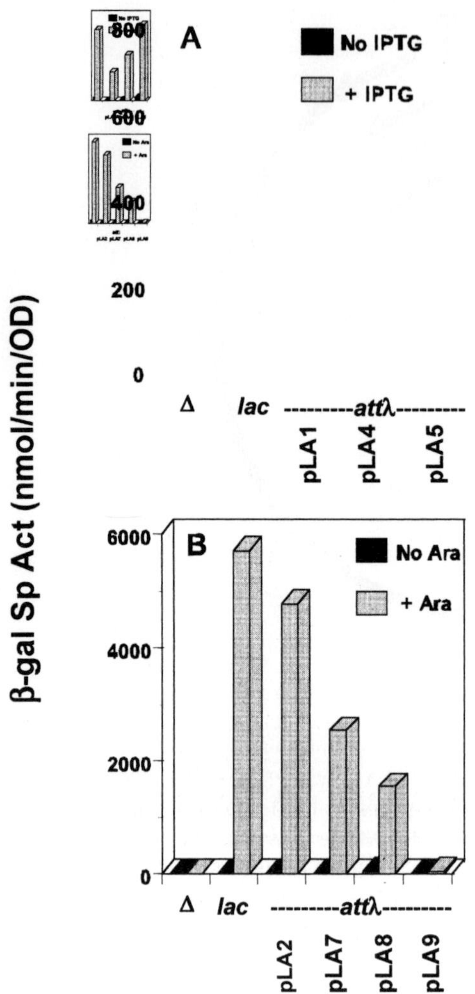

Figure 8   Regulation of LacI- and AraC-controlled *lacZ* fusions. Panel A. Δ, BW22653 (Δ*lacZ*); *lac*, BW25993 (*lacI*q *lacZ*+), pLA1, BW26577 (UV5i- *lacZ*+); pLA4, BW26654 (*lacP*- *lacZ*+); and pLA5, BW26656 (uv5-*lacZ*+) were assayed following 16-h growth in 0.06% glucose MOPS with or without 1 m*M* IPTG. BW25993 is described in Table 1. BW26577, BW26654, and BW26656 are integrants of BW22653 carrying the respective CRIM plasmid. Panel B. Δ, BW25113 (Δ*lacZ*); *lac*, BW22831 (*araBp3-lacZ*+; this strain was created by recombining the *araBp3* fusion from pAH31 onto the chromosome by homologous recombination in place of *lacP* [26]); pLA2, BW26661 (*araBp7-lacZ*+); pLA7, BW26663 (*araBp8-lacZ*+); pLA8, BW26662 (*araBp6-lacZ*+); and pLA9, BW26664 (*araBp3-lacZ*+) were assayed following 16-h growth in 0.1% glycerol MOPS with or without 1 m*M* L-arabinose. The induced level in BW26664 is 2% of the level in BW22831 and ca. 40-fold higher than its uninduced level. BW22831 is described in Table 1. BW26661, BW26662, BW26663, and BW26664 are integrants of BW25113 carrying the respective CRIM plasmid. See text.

recombined onto the chromosome in front of the *lac* operon (Fig. 8; [26]). Yet, it is highly inducible (ca. 40-fold) by arabinose, in agreement with studies showing that a similar CRIM plasmid with an *araBp3-phoB*+ fusion displays an arabinose-dependent PhoB phenotype [26]. The low expression level of *araBp3* in CRIM plasmids is attributable to the fortuitous creation of an artificial AraC' fusion protein that interferes with normal inducibility, as described in the legend of Figure 8.

Several reporter systems are now available for constructing fusions. Major advantages of *lacZ* fusions are the availability of alternative substrates, ease of assay, and, perhaps most of all, genetic selections [48]. X-gal is clearly the most sensitive and widely used substrate for detecting *lacZ* activity. X-gal itself is colorless. Upon cleavage of X-gal by β-galactosidase, an indigogenic product is formed that is spontaneously oxidized in air to an insoluble blue pigment [49]. Because its uptake does not require LacY, X-gal is an especially useful indicator of β-galactosidase in most cells. However, its sensitivity also makes it less useful in genetic selections that require differentiating mutants with elevated or decreased *lacZ* expression levels.

In genetic selections, it is generally more useful to use indicator media requiring both LacZ and LacY functions such as lactose MacConkey and lactose tetrazolium. On these media, color differences result from acid production (MacConkey) or redox chemistry (tetrazolium). The preparation and use of these and other Lac-indicator media are described elsewhere [50, 51]. Importantly, unlike media containing X-gal, these can be used to differentiate mutants showing relatively small (10-fold or less) differences in *lac* expression levels. To facilitate the use of *lacZ* reporter CRIM plasmids in genetic selections requiring LacY function, we use Δ*lacZ* mutants in which *lacY* is induced by IPTG under LacI control (Fig. 9).

ΔlacZ1229::cat lacP-lacY fusion

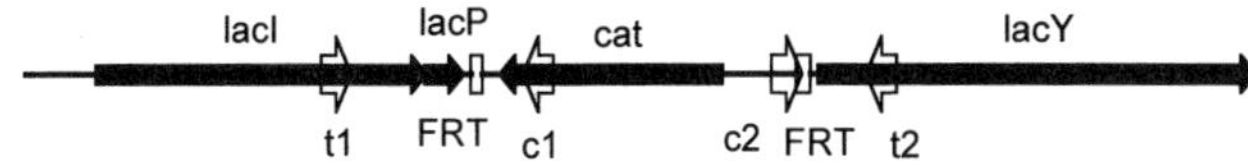

ΔlacZ1230 lacP-lacY fusion

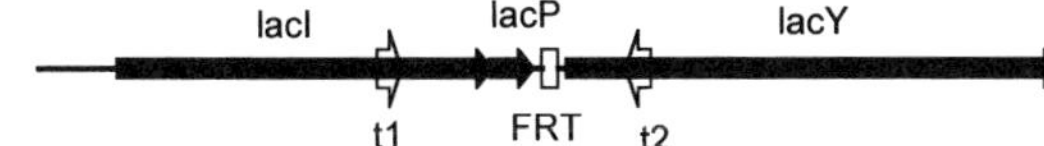

Figure 9 Structures of the Δ*lacZ1229::cat* and Δ*lacZ1230 lacP-lacY* fusions. Strains carrying these fusions were constructed as described in Table 1. The test (t1 and t2) and *cat* (c1 and c2) primers were used to verify the structures by using a standard PCR strategy [27]. The primers t1 and t2 were also used to sequence the Δ*lacZ1230 lacP-lacY* fusion (Genbank accession number AF548059). See text.

6   Remarks and conclusions

In this chapter, we describe several *lacZ* reporter CRIM plasmids for easy construction of specific or random *lacZ* fusions. These have been especially useful not only in studying gene expression in *E. coli* and *S. typhimurium* but also in studying genes from diverse bacteria, including gram-negative and -positive cells [23, 26, 32, 52]. Indeed, by using *E. coli* as a surrogate host and CRIM plasmids encoding various regulatory proteins, we have reconstituted functional regulatory circuits from gram-positive bacteria in *E. coli*. For example, we used a combination of CRIM plasmids and *lacZ* reporter fusions to examine two-component regulatory networks governing vancomycin-resistance systems from enterococci in *E. coli* [52, 53]. Furthermore, we extended these studies by isolating and characterizing mutants of the respective regulatory genes by using *lacZ* fusions similar to those described here (unpublished data).

It should be straightforward to extend these *lacZ* and other CRIM plasmids to other bacteria. We already showed that *attP22* and *attλ* CRIM plasmids integrate into the appropriate *attB* sites of S. *typhimurium* [23]. It should therefore be feasible to use CRIM plasmids not only to construct specific *lacZ* fusions to examine genetic regulatory mechanisms in *S. typhimurium* but also to make random fusion libraries to screen in *S. typhimurium*. Once fusions are found, the respective *lacZ* fusion CRIM plasmid could then be retrieved essentially as done in *E. coli*. This could be done by using a *S. typhimurium galE* mutant, which, unlike wild-type *S. typhimurium,* is sensitive to phage P1*kc* [54]. Alternatively, one could use phage P22 transduction in much the same way by using a recipient with the respective CRIM helper plasmid that is either a *pir⁺ S. typhimurium* or a *pir⁺ E. coli* carrying a cosmid conferring P22 sensitivity [55].

Several F' factors carrying various *attB* sites also exist [56]. By integrating *lacZ* fusion CRIM plasmids into an appropriate F' factor, it would be simple to transfer the fusion into other hosts by bacterial conjugation, a much more efficient process than phage transduction or DNA transformation. These constructs also could be used to introduce *lacZ* fusion CRIM plasmids into closely related bacteria that may lack particular *attB* sites. Also, unlike F' factors carrying *lacZ* fusion phages, F' factors harboring *lacZ* reporter CRIM plasmids would not undergo zygotic induction upon conjugal transfer. Alternatively, low-copy-number, broad-host-range plasmids with an *attB* site could be engineered. To do this, it would be especially advantageous to use a plasmid that also can be transferred by conjugation. Studies using such approaches with *lacZ* reporter CRIM plasmids to extend CRIM technology to diverse bacteria are underway. In conjunction with efficient means for gene disruption [27], *lacZ* reporter and other CRIM plasmids [23] provide new tools for gene analysis on a genome-wide scale.

Acknowledgements

We thank other lab members both past and present who have contributed to our development of various CRIM plasmids. Research has been supported by the U.S. Public Health Service National Institutes of Health (GM57695 and GM62449) and U.S. National Science Foundation (MCB-0110656).

Further reading

Beckwith J (2000) The all purpose gene fusion. *Methods Enzymol* 326: 3–7
Datsenko KA and Wanner BL (2000) One-step inactivation of chromosomal genes in *Escherichia coli* K-12 using PCR products. *Proc Natl Acad Sci USA* 97: 6640–6645

Haldimann A, Prahalad MK, Fisher SL, et al. (1996) Altered recognition mutants of the response regulator PhoB: a new genetic strategy for studying protein-protein interactions. *Proc Natl Acad Sci USA* 93: 14361–14366

Haldimann A and Wanner BL (2001) Conditional-replication, integration, excision, and retrieval plasmid-host systems for gene structure-function studies in bacteria. *J Bacteriol* 183: 6384–6393

References

1 Jacob F, Ullmann A, Monod J (1965) Délétions fusionnant l'opéron lactose et un opéron purine chez *Escherichia coli*. *J Mol Biol* 31: 704–719

2 Silhavy TJ, Beckwith JR (1985) Uses of *lac* fusions for the study of biological problems. *Microbiol Rev* 49: 398–418

3 Bier E, Vaessin H, Shepherd S, et al. (1989) Searching for pattern and mutation in the Drosophila genome with a P-*lacZ* vector. *Genes Dev* 3: 1273–1287

4 Fire A, Harrison SW, Dixon D (1990) A modular set of *lacZ* fusion vectors for studying gene expression in *Caenorhabditis elegans*. *Gene* 93: 189–198

5 Mansour SL, Thomas KR, Deng CX, Capecchi MR (1990) Introduction of a *lacZ* reporter gene into the mouse int-2 locus by homologous recombination *Proc Natl Acad Sci USA* 87: 7688–7692

6 Beckwith JR (1963) Restoration of operon activity by suppressors. *Biochim Biophys Acta* 76: 162–164

7 Miller JH, Reznikoff WS, Silverstone AE et al. (1970) Fusions of the *lac* and *trp* regions of the *Escherichia coli* chromosome. *J Bacteriol* 104: 1273–1279

8 Casadaban MJ (1975) Fusion of the *Escherichia coli lac* genes to the *ara* promoter: A general technique using bacterial Mu-1 insertions. *Proc Natl Acad Sci USA* 72: 809–713

9 Casadaban MJ, Cohen SN (1979) Lactose genes fused to exogenous promoters in one step using a Mu-*lac* bacteriophage: *In vivo* probe for transcriptional control sequences. *Proc Natl Acad Sci USA* 76: 4530–4533

10 Casadaban MJ, Chou J, Cohen SN (1980) *In vitro* gene fusions that join an enzymatically active b-galactosidase segment to amino-terminal fragments of exogenous proteins: *Escherichia coli* plasmid vectors for the detection and cloning of translational initiation signals. *J Bacteriol* 143: 971–980

11 Simons RW, Houman F, Kleckner N (1987) Improved single and multicopy *lac*-based cloning vectors for protein and operon fusions. *Gene* 53: 85–96

12 Linn T, St. Pierre R (1990) Improved vector system for constructing transcriptional fusions that ensures independent translation of *lacZ*. *J Bacteriol* 172: 1077–1084

13 Ostrow KS, Silhavy TJ, Garrett S (1986) *cis*-acting sites required for osmoregulation of *ompF* expression in *Escherichia coli* K-12. *J Bacteriol* 168: 1165–1171

14 Way JC, Davis MA, Morisato D et al. (1984) New Tn*10* derivatives for transposon mutagenesis and for construction of *lacZ* operon fusions by transposition. *Gene* 32: 369–379

15 De Lorenzo V, Herrero M, Jakubzik U, Timmis KN (1990) Mini-Tn*5* transposon derivatives for insertion mutagenesis, promoter probing, and chromosomal insertion of cloned DNA in gram-negative eubacteria. *J Bacteriol* 172: 6568–6572

16 Huisman O, Raymond W, Froehlich KU et al. (1987) A Tn10-lacZ-kanR-URA3 gene fusion transposon for insertion mutagenesis and fusion anlysis of yeast and bacterial genes. *Genetics* 116: 191–199

17 Manoil C (1991) Analysis of membrane protein topology using alkaline phospha-

tase and b-galactosidase gene fusions. *Methods Cell Biol* 34: 61–75

18 Wilmes-Riesenberg MR, Wanner BL (1992) Tn*phoA* and Tn*phoA*' elements for making and switching fusions for study of transcription, translation, and cell surface localization. *J Bacteriol* 174: 4558–4575

19 Biery MC, Stewart FJ, Stellwagen AE et al. (2000) A simple *in vitro* Tn7-based transposition system with low target site selectivity for genome and gene analysis. *Nucleic Acids Res* 28: 1067–1077

20 Goryshin IY, Jendrisak J, Hoffman LM et al. (2000) Insertional transposon mutagenesis by electroporation of released Tn5 transposition complexes. *Nat Biotechnol* 18: 97–100

21 Griffin TJ, Parsons L, Leschziner AE et al. (1999) *In vitro* transposition of Tn*552*: a tool for DNA sequencing and mutagenesis. *Nucleic Acids Res* 27: 3859–3865

22 Akerley BJ, Rubin EJ, Camilli A et al. (1998) Systematic identification of essential genes by *in vitro mariner* mutagenesis. *Proc Natl Acad Sci USA* 95: 8927–8932

23 Haldimann A, Wanner BL (2001) Conditional-replication, integration, excision, and retrieval plasmid-host systems for gene structure-function studies in bacteria. *J Bacteriol* 183: 6384–6393

24 Majumder K, Choudhury S, Bhatnagar RK (1994) Recombinant enrichment by exploitation of restriction sites with interrupted palindromes: Design, synthesis and incorporation of zero-background linkers in cloning and expression vectors. *Gene* 151: 147–151

25 Lessard IAD, Pratt SD, McCafferty DG et al. (1998) Homologs of the vancomycin resistance D-ala-D-ala dipeptidase VanX in *Streptomyces toyocaensis*, *Escherichia coli*, and *Synechocystis*: Attributes of catalytic efficiency, stereoselectivity, and regulation with implications for function. *Chemistry & Biology* 5: 489–504

26 Haldimann A, Daniels LL, Wanner BL (1998) Use of new methods for construction of tightly regulated arabinose and rhamnose promoter fusions in studies of the *Escherichia coli* phosphate regulon. *J Bacteriol* 180: 1277–1286

27 Datsenko KA, Wanner BL (2000) One-step inactivation of chromosomal genes in *Escherichia coli* K-12 using PCR products. *Proc Natl Acad Sci USA* 97: 6640–6645

28 Khlebnikov A, Datsenko KA, Skaug T et al. (2001) Homogeneous expression of the P_{BAD} promoter in *Escherichia coli* by constitutive expression of the low-affinity high-capacity AraE transporter. *Microbiol* 147: 3241–3247

29 Jensen KF (1993) The *Escherichia coli* K-12 "wild types" W3110 and MG1655 have an *rph* frameshift mutation that leads to pyrimidine starvation due to low *pyrE* expression levels. *J Bacteriol* 175: 3401–3407

30 Wanner BL (1994) Gene expression in bacteria using Tn*phoA* and Tn*phoA*' elements to make and switch *phoA* gene, *lacZ* (op), and *lacZ* (pr) fusions. In: KW Adolph (ed): *Methods in molecular genetics*, Vol. 3. Academic Press, Orlando, 291–310

31 Hanahan D (1983) Studies on transformation of *Escherichia coli* with plasmids. *J Mol Biol* 166: 557–580

32 Haldimann A, Prahalad MK, Fisher SL et al. (1996) Altered recognition mutants of the response regulator PhoB: a new genetic strategy for studying protein-protein interactions. *Proc Natl Acad Sci USA* 93: 14361–14366

33 Agrawal DK, Wanner BL (1990) A *phoA* structural gene mutation that conditionally affects formation of the enzyme bacterial alkaline phosphatase. *J Bacteriol* 172: 3180–3190

34 Metcalf WW, Jiang W, Wanner BL (1994) Use of the *rep* technique for allele replacement to construct new *Escherichia coli* hosts for maintenance of R6Kg origin plasmids at different copy numbers. *Gene* 138: 1–7

35 Ausubel FM, Brent R, Kingston RE et al. (2002) *Current protocols in molecular biology*. John Wiley & Sons, New York

36 Beckwith JR, Signer ER (1966) Transposition of the *Lac* Region of *Escherichia*

coli. I. Inversion of the *Lac* Operon and Transduction of *Lac* by f80. *J Mol Biol* 19: 254–265

37 Reznikoff WS, Miller JH, Scaife JG, Beckwith JR (1969) A mechanism for repressor action. *J Mol Biol* 43: 201–213

38 Sarthy A, Fowler A, Zabin I, Beckwith J (1979) Use of gene fusions to determine a partial signal sequence of alkaline phosphatase. *J Bacteriol* 139: 932–939

39 Kenyon CJ, Walker GC (1980) DNA-damaging agents stimulate gene expression at specific loci in *Escherichia coli*. *Proc Natl Acad Sci USA* 77: 2819–2823

40 Wanner BL, Wieder S, McSharry R (1981). Use of bacteriophage transposon Mu *d1* to determine the orientation for three *proC*-linked phosphate-starvation-inducible (*psi*) genes in *Escherichia coli* K-12. *J Bacteriol* 146: 93–101

41 Mahan MJ, Slauch JM, Mekalanos JJ (1993) Selection of bacterial virulence genes that are specifically induced in host tissues. *Science* 259: 686–688

42 Brown S (1987) Mutations in the gene for EF-G reduce the requirement for 4.5S RNA in the growth of E. coli. *Cell* 49: 825–833

43 Guzman LM, Belin D, Carson MJ, Beckwith J (1995) Tight regulation, modulation, and high-level expression by vectors containing the arabinose P_{BAD} promoter. *J Bacteriol* 177: 4121–4130

44 Müller J, Oehler S, Müller-Hill B (1996) Repression of *lac* promoter as a function of distance, phase and quality of an auxiliary *lac* operator. *J Mol Biol* 257: 21–29

45 Yanisch-Perron C, Vieira J, Messing J (1985) Improved M13 phage cloning vectors and host strains: nucleotide sequences of the M13mp18 and pUC19 vectors. *Gene* 33: 103–119

46 Noel RJ, Reznikoff WS (2000) Structural studies of *lac*UV5-RNA polymerase interactions *in vitro* - Ethylation interference and missing nucleoside analysis. *J Biol Chem* 275: 7708–7712

47 Dunn T, Hahn S, Ogden S, Schleif R (1984) An *araBAD* operator at -280

base pairs that is required for pBAD repression: addition of DNA helical turns between the operator and promoter cylindrically hinders repression. *Proc Natl Acad Sci USA* 81: 5017–5020

48 Beckwith JR (1970) Lac: The genetic system. In: JR Beckwith, D Zipser (eds*): The Lactose Operon*. Cold Spring Harbor Laboratory, Cold Spring Harbor, New York, 5–26

49 Horwitz JP, Chua J, Curby RJ et al. (1964) Substrates for cytochemical demonstration of enzyme activity. I. Some substituted 3-indolyl-β-D-glycopyranosides. *J Med Chem* 7: 574

50 Miller JH (1972) *Experiments in molecular genetics*, Cold Spring Harbor Laboratory, Cold Spring Harbor, New York

51 Miller JH (1992) *A short course in bacterial genetics: a laboratory manual and handbook for Escherichia coli and related bacteria*. Cold Spring Harbor Laboratory Press, Plainview, NY

52 Haldimann A, Fisher SL, Daniels LL et al. (1997) Transcriptional regulation of the *Enterococcus faecium* BM4147 vancomycin resistance gene cluster by the VanS-VanR two-component regulatory system in *Escherichia coli*. *J Bacteriol* 179: 5903–5913

53 Silva JC, Haldimann A, Prahalad MK et al. (1998) *In vivo* characterization of the type A and B vancomycin-resistant enterococci (VRE) VanRS two-component systems in *Escherichia coli*: A nonpathogenic model for studying the VRE signal transduction pathways. *Proc Natl Acad Sci USA* 95: 11951-11956

54 Mojica T (1975) Transduction by phage P1CM clr-100 in *Salmonella typhimurium*. *Mol Gen Genet* 138: 113–126

55 Neal BL, Brown PK, Reeves PR (1993) Use of *Salmonella* phage P22 for transduction in *Escherichia coli*. *J Bacteriol* 175: 7115–7118

56 Low KB (1972) *Escherichia coli* K-12 F-prime factors, old and new. *Bacteriol Rev* 36: 587–607

# Genetic Footprinting for Bacterial Functional Genomics

Scott S. Walker, Chad Houseweart and Teresa J. Kenney

Contents

1   Introduction

As more complete bacterial genome sequences become available, the need for broad, genomic-scale analytical methods becomes increasingly acute. Moreover, with a sizable portion of microbial genomes encoding proteins of unknown function, tools that suggest gene function or elucidate relationships among

Methods and Tools in Biosciences and Medicine
Prokaryotic Genomics, ed. by M. Blot
© 2003 Birkhäuser Verlag Basel/Switzerland

genes are crucial for any genomic-scale evaluation. While genomics tools may be central to a basic scientist's interests in a model organism, they will also be critical to the applied scientist's examination of a pathogen for targets for therapeutic intervention.

A major consideration in functional studies is to determine the importance of a gene for cell viability. Classically, in bacteria, gene essentiality was inferred from isolating a conditional mutant (e. g., temperature-sensitive or regulated expression) or from the inability to generate a targeted gene disruption by homologous recombination. Although useful, these methods are difficult to apply on a genome-wide scale. Recently, three related methods have been developed to apply gene-disruption technology to the interrogation of an entire bacterial genome [1–3].

The central theme of these three gene-disruption methods is transposon-mediated, insertional mutagenesis. However, they differ in the method of transposon delivery. Two of the methods, GAMBIT [1] and Ty-1 genome scanning [2], rely on *in vitro* transposon mutagenesis of genomic DNA (or PCR products) and transformation of the host bacteria with the mutagenized sample. Transformation is followed by selection for the transposon and detection of insertions in a gene of interest. These methods are limited to organisms amenable to linear transformation and efficient recombination. The third method, genetic footprinting, uses *in vivo* plasmid- or phage-delivered transposon-mediated mutagenesis [3–5]. Originally developed for use in the yeast *Saccharomyces cerevisiae* [4], genetic footprinting has been adapted to the bacterium *Escherichia coli* [3].

Genetic footprinting, as depicted in Figure 1, is a multistep process involving transient induction of transposon mobilization, outgrowth of the mutagenized cell population, and examination of the fate of cells bearing an insertion mutation within a gene of interest. To measure the importance of a particular gene to cell viability, gene and transposon-specific PCR primers are used to detect insertions within the gene of interest in the initial cell population (T0) and in the outgrowth population sampled at various stages (e. g., T15, T30, or T45 [15, 30, or 45 doublings of the initial population]). Gene essentiality is inferred from the appearance of PCR products indicative of transposons within the gene of interest in the initial cell population and from the subsequent loss of PCR products generated from cells collected from the outgrowth culture, creating a "footprint" (Fig. 1). Analysis of the rate of loss of PCR products corresponding to insertions in a specific gene is indicative of the growth rate of these mutants relative to the cell population [5]. When the data are analyzed in this way, footprinting can expose even subtle effects of gene loss not suggested from traditional gene-disruption and phenotypic analysis [4, 5].

The design of an *in vivo* system for bacterial genetic footprinting should include the following features: an efficient and random transposable element, a transposase dependent upon a regulated expression system, and a regulated plasmid replicon to further decrease transposase expression [3, 6]. Alternatively, using a replication-defective, phage-based method obviates the requirement for a regu-

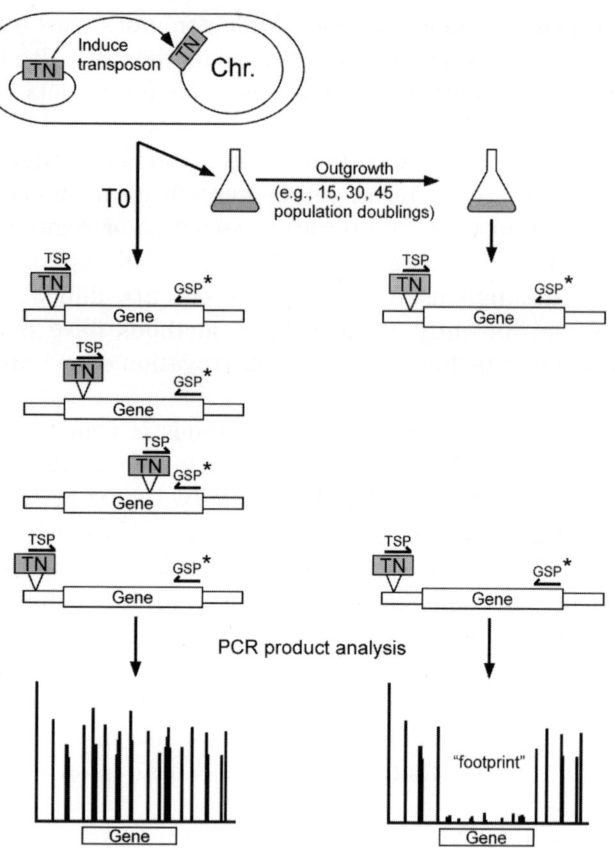

Figure 1 Bacterial genetic footprinting. Schematic depiction of the major aspects of genetic footprinting. The key steps of genetic footprinting are transient transposon induction, initial sampling, outgrowth and sampling, and PCR product analysis. TSP: transposon-specific primer, GSP*: fluorescently labeled, gene-specific primer.

lated replicon. In bacteria, genetic footprinting is complicated by the poly-cistronic nature of prokaryotic chromosomal architecture. Polarity issues within operons may be abrogated by including a strong, outwardly oriented promoter within the transposable element [3]. With these considerations appropriately addressed, genetic footprinting is a robust analytical method for examining the role of numerous genes in a cell under a variety of growth conditions.

2   Materials

2.1   Microbial growth media

Medium 1: L broth
L broth contains 20 g tryptone, 10 g yeast extract, and 7 g NaCl per liter, and M9 minimal media contains 0.2% glucose, 1 mM $MgSO_4$, 1 µg/ml thiamine-HCl, and 0.1 mM $CaCl_2$ [7]. Antibiotics should be used at the following final concentra-

tions: ampicillin (Amp) at 100 µg/ml, kanamycin (Kan) at 30 µg/ml, and chloramphenicol (Cam) at 30 µg/ml. Isopropylthiogalactopyranoside (IPTG) is used at 1 mM final concentration, unless noted otherwise.

Medium 2: TBMM broth
TBMM broth contains 10 g/l tryptone, 5 g/l NaCl, 0.2% maltose, 10 mM $MgSO_4$, and 1 µg/ml thiamine, as described [6].

Medium 3: LB broth
LB broth contains 10 g tryptone, 5 g yeast extract, and 10 g NaCl per liter.

2.2 Solutions

Solution 1: PCR buffer
PCR buffer contains 10 mM Tris(hydroxymethyl)aminomethane-hydrochloride (Tris-HCl) pH 8.3, 50 mM KCl, 1.5 mM $MgCl_2$, and 0.001% gelatin.

Solution 2: Loading solution
Loading solution contains 100 mg/ml blue dextran, GeneScan™ 2500 Tamara gene standards (Perkin-Elmer), and deionized formamide in a 1:1:5 ratio.

2.3 Strains, plasmids, and bacteriophage

E. coli strains MG1655, W3110, and W3110 T7 RNAP have been described [3, 8]. The relevant features of the transposon-delivery plasmid (pGT-G69), inducible replicon and transposase, and a transposon with an outwardly oriented promoter are described in Hare et al. (2001) [3]. Construction of the bacteriophage λ transposon delivery-vehicle, λHS, also has been described [3]. The *E. coli murA* and *murB* genes were cloned into the plasmid pET-29a(+) as described [3].

2.4 PCR primer design and synthesis

Useful criteria for oligonucleotide primer design have been described previously [4]. Briefly, gene-specific primers should be chosen with aid of a primer analysis program, (e. g., Oligo 4.06, National Biosciences, Plymouth, MN) such that they anneal within 300 to 1500 bp of the 5' terminus of each gene and have a calculated melting temperature (T_m) of 53–63 ºC. Synthesize gene-specific primers (usually 24mers) with a covalently attached 6-FAM fluorescent group at the 5' terminus (Research Genetics, Inc., Huntsville, AL). One of three previously described unlabeled transposon-specific primers was used in the examples shown here [3]. Two of the primers are complementary to both IS*10*

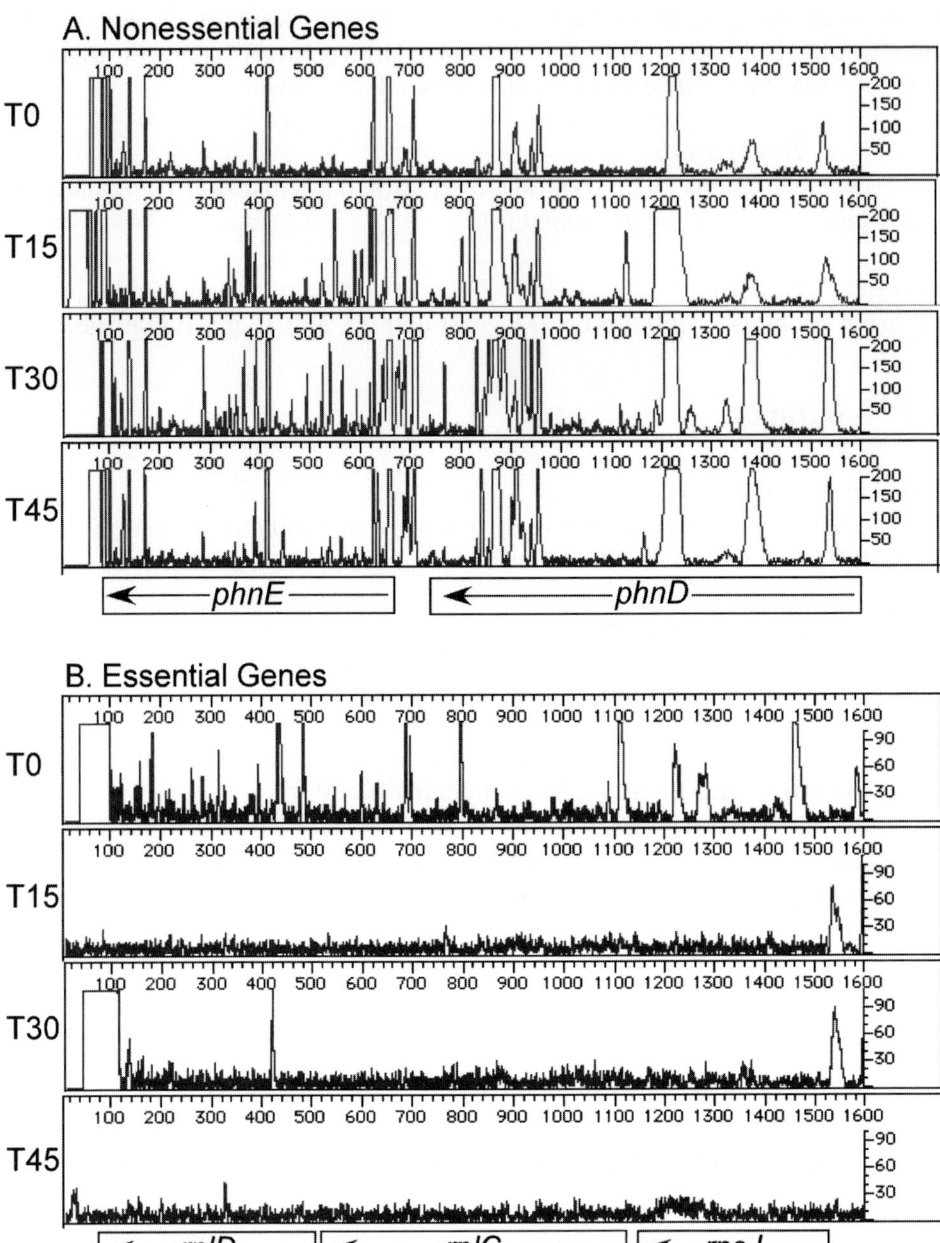

Figure 2   Plasmid based genetic footprinting. (A) Example of the genetic footprint (Tn10 insertion profile) of two nonessential genes. (B) Example of the genetic footprint (Tn10 insertion profile) of three essential genes. T0 is the point at which transposition is repressed, and T15, T30, and T45 represent the number of population doublings following T0. The histograms depict the abundance *versus* length of PCR products generated from a transposon-specific and gene-specific primer pair. The regions corresponding to the analyzed ORFs are denoted by boxes. The arrows indicate the direction of transcription. (Adapted from Hare et al. (2001) used with permission from the American Society for Microbiology.)

elements of the pNK2887-derived mini-Tn*10* [3, 6]. The third transposon-specific primer anneals to a region asymmetrically located within the transposon.

2.5 Equipment and software

Automated sequencer
Use the Prism DNA sequencer model 377XL from Applied Biosystems Inc. (Perkin-Elmer) or an appropriate substitute.

Software
Sequencer-associated software version 2.0 and GeneScan™ version 2.0.2 can be used for initial analysis. Subsequent analysis and display (e. g., see Fig. 2) is done with the ABI Genotyper™ software package, version 2.0.

3   Methods

Protocol 1   Plasmid-based footprinting: Transposase induction and selection for Tn*10* transposition: The T0 population

1. Grow *E. coli* strain MG1655 *rpsL* (pGT-G69) or another suitable host in 50 ml L broth (Medium 1, containing Amp and IPTG) until the cells reach an optical density at 600 nm (OD_{600}) of 0.8 ($2–5 \times 10^8$ cells/ml).
2. Add the 50 mL culture to 1 l of L broth supplemented with Amp and IPTG and incubated at 37 °C until the cells reach $OD_{600} = 0.8$ (this represents a growth time of approximately 4 h). Repeat this cycle three additional times for a total of four growth cycles.
3. To obtain sufficient T0 DNA for footprinting on a genomic scale, at the end of the fourth growth cycle, add 500 ml of culture to a fermentation vessel containing 10 l of L broth and grow until the $OD_{600} = 0.8$. In total, the five growth cycles represent approximately 25 population doublings.
4. Harvest cells by centrifugation (5000 rpm), resuspend in L broth containing 15% glycerol at a concentration of 40:1, dispense in 2 ml fractions, and store at –80 °C. The DNA and cells obtained from these frozen samples is designated as the T0 sample.

Protocol 2 Outgrowth of mutagenized cell populations: Isolation of the T15, T30, and T45 populations

1. Inoculate 500 ml of L broth lacking IPTG with a 2 ml aliquot of frozen cells (1.6–8 x 10^{10} cells) from the T0 culture. Incubate at 37 °C until an OD_{600} = 0.8 is reached.
2. Transfer 50 ml of the culture to 1 l of rich medium and grow for 45 population doublings by repeating this step 8–9 times.
3. During outgrowth, withdraw 50 ml samples at T15 and T30 and treat as described for T45 (below).
4. At T45, harvest the culture and resuspend in 15% glycerol at a concentration of 40:1. Freeze aliquots for subsequent isolation of chromosomal DNA.
5. For alternate outgrowth conditions, the growth procedure described above can be repeated with different media including minimal media and media supplimented with or lacking particular nutrients or chemicals (see Remarks and conclusions, below).

Protocol 3 Infection and analysis for bacteriophage λ-based transposon delivery

Infect the bacterial cells containing a tagged gene to be tested for complementation with λHS as follows:
1. Grow the transformed W3110 T7 RNAP strain in 10 ml TBMM broth (Medium 2) plus 35 µg/ml Kan, at 37 °C for 16 h. Harvest cells and resuspend in 1 ml LB broth (Medium 3).
2. Infect these cells with 0.5 ml λ phage lysate (multiplicity of infection = 1) and incubate at room temperature for 15 min and then at 37 °C for 15 min.
3. Transfer cells to a 250 ml flask containing 50 ml LB broth, supplemented with 50 mM sodium citrate and 50 µM IPTG. Shake for 1 h at 37 °C.
4. Harvest cells and transfer to 1 l of LB broth containing 5 mM sodium citrate, 35 µg/ml Cam, 35 µg/ml Kan, and 50 µM IPTG. Incubate culture with shaking at 30 °C for 14–16 hours or until the OD_{600} = 1.0 to 1.4.
5. Harvest cells from 100 ml of the culture, resuspend in 2 ml LB broth plus 10% glycerol, freeze in a dry ice-ethanol bath and store at –80 °C.

Protocol 4 Genomic DNA isolation

1. Isolate bacterial chromosomal DNA using the Qiagen Genomic Tip-500 according to the manufacturer's instructions (Qiagen, Chatsworth, CA).
2. Dissolve DNA in 10 mM Tris-HCl (pH 8.0).

Protocol 5   Polymerase chain reaction

Conditions for PCR are similar to those described by Smith et al. [4, 5].
1. Set up a polymerase chain reaction containing 0.5 µg template DNA, 0.5 µM each primer, 250 µM each deoxynucleoside triphosphate, PCR buffer (Solution 1), and 2 units of *Taq* DNA polymerase (Perkin-Elmer, Foster City, CA) in a final volume of 50 µl.
2. Carry out the reaction at 94 °C for 1 min; 92 °C, 30 s; 67 °C, 45 s; 72 °C, 2 min (10 cycles); 92 °C, 30 s; 62 °C, 45 s; 72 °C, 2 min (20 cycles); 72 °C, 3 min.

Protocol 6   PCR product analysis

1. Add 3 µl of the PCR products to 3 µl of loading solution (Solution 2).
2. Denature samples for 3 min at 100 °C, place immediately in an ice-water bath, and load (1.5–3 µl) onto a 5% polyacrylamide gel (Long Ranger™, BioWhittaker, Rockland, ME) containing 6 M urea and 1 x TBE buffer (89 mM Tris base, 89 mM boric acid, 2 mM EDTA, pH 8.0).
3. Electrophoresis and analysis of the PCR products are performed using the ABI 377 Prism DNA sequencer and associated software. Size and distribution of PCR products are further analyzed with the ABI Genotyper™ software package. In Genotyper™, adjust the vertical scale such that the strong peaks approach full scale.

4   Troubleshooting

- *No PCR product:* Removal of EDTA (ethylamine diamine tetraacetic acid) from the genomic DNA preparation may significantly increase PCR product yield.
- *Insufficient PCR product:* If initial PCR conditions and removal of EDTA do not produce a robust product, the annealing temperatures can be reduced to 62 °C in the first 10 cycles and to 58 °C for the next 22 cycles. The total number of amplification cycles also may be increased to 35 to improve yield. Further improvement in product yield may be achieved by replacing the Taq polymerase with either Platinum Taq or Titanium Taq (Clontech, Palo Alto, CA).
- *Primer resynthesis:* In some cases where no signal or a particularly strong peak is evident, primer resynthesis may be useful. Primer length may be increased to 27 nucleotides to increase DNA duplex T_m. Considerations for primer redesign include open reading frame (ORF) size (see below) and redundancy of a chosen sequence within the genome. Also, switching the location of the primer from the 3' end of the gene to the 5' end of the gene can be helpful.

- *Small genes:* While footprinting small genes (less than 400 bp) may be problematic, they can sometimes be successfully analyzed by choosing a primer within an adjacent gene or untranscribed region. However, interpretation of results may be complicated by the lack of sufficient transposon insertions within the small gene of interest.
- *Interpretation of data:* When results appear ambiguous, repeating the reactions with alternate transposon specific primers may clarify data analysis.
- *Outgrowth conditions:* Conditions for outgrowth may need to be optimized based on bacterial growth rate under a given media condition and the number of generations needed to clear the transposon delivery system.
- *General controls:* Tests of known nonessential and essential genes are crucial to optimize any experimental growth or analytical (e. g., PCR) conditions.

5   Applications

5.1   Example 1: Gene essentiality testing

Genetic footprinting is a genomic-scale method useful for gauging the importance of a gene of interest for the competitive growth capacity of the cell in a mixed culture [3–5]. When carried out in rich-culture conditions, genetic footprinting can identify the core set of essential genes in the cell while avoiding simple auxotrophies. Figure 2 A shows an analysis of two nonessential genes, *phnE* and *phnD*, involved in phosphonate transport [9]. There is no significant drop in the abundance of PCR products corresponding to insertions in either of these two genes, even after 45 generations of competitive growth. On the other hand, the pattern seen when analyzing the essential ribosomal subunits, *rplD, rplC,* and *rpsJ*, suggests a profound effect on the viability of cells with mutations within these genes even at the first time point (Fig. 2B).

5.2   Example 2: Cloned gene complementation testing

In addition to gene-essentiality testing in rich or defined conditional growth media, genetic footprinting also can be applied to complementation analysis of a cloned gene [3]. Such an analysis requires that the complementing plasmid is compatible with the transposon delivery plasmid. A replication-defective, bacteriophage-based system obviates this issue. Figure 3 shows an example of bacteriophage lambda-mediated transposon delivery. In this example, a plasmid containing the essential gene, *murB*, renders the chromosomal copy of *murB* nonessential, whereas a plasmid containing *murA* does not.

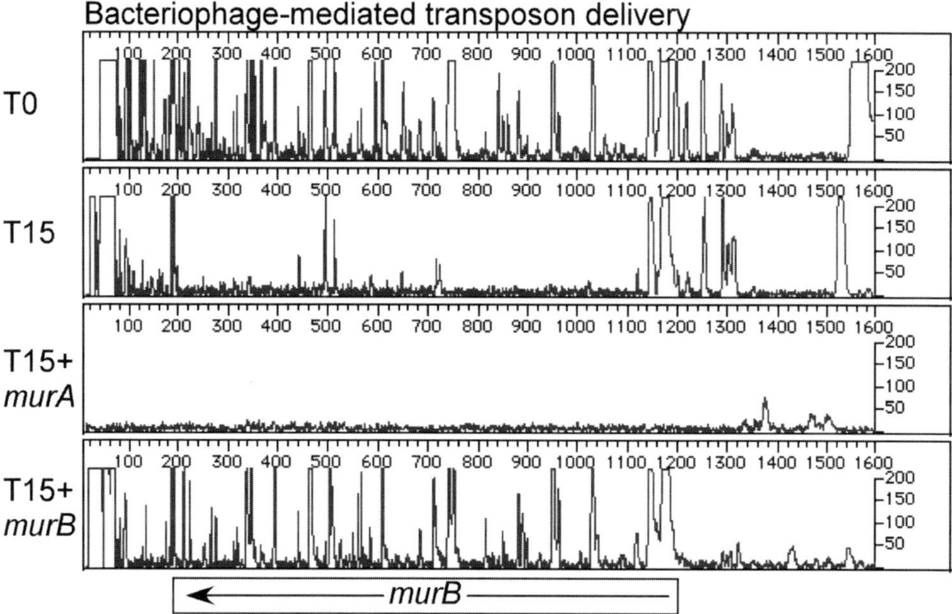

Figure 3 Bacteriophage lambda-mediated genetic footprinting. Example of genetic footprinting and cloned gene complementation testing using bacteriophage-based transposon delivery. The gene-specific (*murB*) primer used in conjunction with a transposon-specific primer anneals outside of the region of *murB* in the expression plasmid [3]. (Adapted from Hare et al. (2001) used with permission from the American Society for Microbiology.)

6 Remarks and conclusions

Genetic footprinting provides a robust means to expose even subtle effects of gene disruption on the competitive growth of a cell in mixed culture. If the goal of the analysis is to identify only the broadly required genes (essential regardless of growth conditions), then those genes that show a strong footprint at the earliest time point should be the focus. When examining a genome, genetic footprinting provides a feasible first-pass method to identify the majority of the essential genes in a bacterial cell. In addition, any follow-up, targeted gene-disruption experiments to confirm a result or to produce mutants for further use are greatly facilitated by the reduced number genes to examine (see Further reading).

While essential genes will be identified by genetic footprinting, the ability to manipulate culture conditions and the sensitive nature of this method make genetic footprinting well suited to exposing genes required for viability under defined conditions. These conditions might include nutrient deprivation (e. g., minimal media lacking specific amino acids), a modified culture parameter (e. g., temperature), or application of a chemical stress (e. g., antibiotic treatment). Within any single experiment, once the cells have been grown and sampled, a large number of genes can be examined in parallel.

Acknowledgments

The authors wish to thank Paul M. McNicholas for helpful comments on the manuscript.

Further reading

Examples and references for targeted bacterial gene disruption methods

Link A, Phillips D and Church G (1997) Methods for generating precise deletions and insertions in the genome of wild-type *Escherichia coli*: application to open reading frame characterization. *J Bacteriol* 179: 6228–6237

Posfai G, Kolisnychenko V, Bereczki Z et al. (1999) Markerless gene replacement in *Escherichia coli* stimulated by a double-strand break in the chromosome. *Nucleic Acids Res* 27: 4409–4415

Datsenko K and Wanner B (2000) One-step inactivation of chromosomal genes in *Escherichia coli* K-12 using PCR products. *Proc Natl Acad Sci USA* 97: 6640–6645

Karberg M, Guo H, Zhong J, et al. (2001) Group II introns as controllable gene targeting vectors for genetic manipulation of bacteria. *Nature Biotech* 19: 1162–1167

References

1   Akerley BJ, Rubin EJ, Camilli A, et al. (1998) Systematic identification of essential genes by *in vitro mariner* mutagenesis. *Proc Natl Acad Sci USA* 95: 8927–8932

2   Reich KA, Chovan L, Hessler P (1999) Genome scanning in *Haemophilus influenza* for identification of essential genes. *J Bacteriol* 181: 4961–4968

3   Hare R, Walker S, Dorman T, et al. (2001) Genetic footprinting in bacteria. *J Bacteriology* 183: 1694–1706

4   Smith V, Botstein D, Brown PO (1995) Genetic footprinting: a genomic strategy for determining a gene's function given its sequence. *Proc Natl Acad Sci USA* 92: 6479–6483

5   Smith V, Chou KN, Lashkari D, et al. (1996) Functional analysis of the genes of yeast chromosome V by genetic footprinting. *Science* 274: 2069–2074

6   Kleckner N, Bender J, Gottesman S (1991) Uses of transposons with emphasis on Tn*10*. In: JH Miller (ed): *Methods in enzymology: Bacterial genetic systems*. Academic Press, New York, 139–180

7   Davis RW, Botstein D, Roth JR (1980) *A manual for genetic engineering: Advanced bacterial genetics*. Cold Spring Harbor Laboratory, Cold Spring Harbor, NY

8   Jensen KF (1993) The *Escherichia coli* K12 "wild types" W3110 and MG1655 have an *rph* frameshift mutation that

leads to pyrimidine starvation due to low *pyrE* expression levels. *J Bacteriol* 175: 3401–3407

9   Metcalf WW, Wanner BL (1993) Mutational analysis of an *Escherichia coli* fourteen-gene operon for phosphonate degradation, using Tn*phoA* elements. *J Bacteriol* 175: 3430–3432

Gene Transfer to Plants through Bacterial Vectors

Bruno Tinland

Contents

1 Introduction

For 20 years, research in plant molecular biology and development of plant biotechnology have been tightly coupled with the discovery of *Agrobacterium*-mediated transformation of plants. *Agrobacterium* is a common lab tool for many plant biologists because it has evolved the unique capacity to transfer a piece of its own DNA, the transferred DNA, or T-DNA, into the nuclear genome of plant cells (for review see [1–3]). This property has been extremely useful for the introduction of new genes into plants either for research or applied purposes [4], and for the inactivation of plant genes by insertion mutagenesis [5, 6]. Also, the interaction between *Agrobacterium* and the plant cell itself has provided powerful tools for directly addressing questions of bacteriology and plant biology. For bacteriology this is true in the field of bacterial conjugation, since it appeared that mechanisms involved in the delivery of DNA from bacteria to a plant cell are similar to those involved in the conjugative transfer of DNA between bacteria. In plant biology it is especially true in the domain of

Methods and Tools in Biosciences and Medicine
Prokaryotic Genomics, ed. by M. Blot
© 2003 Birkhäuser Verlag Basel/Switzerland

genetic recombination, because the characteristics of T–DNA-mediated transformation have no known equivalent in any bacterial system.

Agrobacterium is uniquely equipped to fulfill its role as a sophisticated plant parasite. It contains a plasmid called Ti plasmid (Ti = tumor inducing) that encodes most of the major functions required for virulence. One segment of the plasmid, the T-DNA, is designed to be transported to the plant cell, to be integrated into the plant cell nuclear DNA, and to express its genes only in the eukaryotic cellular environment. These genes code for enzymes required for synthesis of new compounds termed opines and for enzymes involved in phytohormone production. The opines serve as a food source for the bacterium, whereas the hormones lead to the neoplastic phenotype of the tumor (or proliferative root development for *A. rhizogenes*). The fact is that none of the T-DNA genes is required for the T-DNA transfer step as such led to the development of the *Agrobacterium*-based transformation technology. T-DNA in *Agrobacterium* is delimited by two 25-bp directs imperfect repeat sequences called borders. Most of the functions necessary for transfer of T-DNA are encoded in the virulence region located on the Ti plasmid; some are located on the chromosome.

Only when a wounded plant is in "useful" distance from a bacterium are the virulence genes induced. Chemical signals emitted by a healing plant wound are, *via* a receptor/activator system (proteins VirA/VirG), converted into transcriptional activation of the virulence genes. Virulence proteins D1 and D2 introduce a specific cut into one strand of the border sequences. Single-stranded DNA molecules derived from the T-DNA (T-strands) are produced and are covalently bound to the VirD2 protein at their 5' extremity. These T-strands are then transported through bacterial pore structures, probably composed of VirB proteins, into plant cells. T-strands are then coated by VirE2 proteins that are also transported to the plant cells. This association between the T-strand and the VirE2 protein is called T-complex.

The T-complex structure protects the integrity of the piece of DNA that has been transferred to the plant cell. It will be targeted to the nucleus *via* nuclear pore recognition and enter it. Finally, the extremities of the T-strand will be ligated to the nuclear DNA following an illegitimate recombination process. Plant enzymes are the main contributors to this recombination process, and, upon duplication of the T-strand, the newly inserted DNA molecule of bacterial origin becomes part of the plant nuclear DNA, i.e., it will duplicate and segregate like the surrounding DNA.

The aim of this paper is to provide a simple experimental setup that allows monitoring of the transfer of the T-DNA from *Agrobacterium* to the plant cell nuclei.

2   Materials

2.1 Solutions

- *MS medium* [7]
- *YEB* [8]

The following solutions are described by Jefferson [9]:
- *GUS extraction buffer*: 50 mM sodium phosphate buffer, pH 7.0; 10 mM DTT; 1 mM Na_2 EDTA; 0.1% sodium lauryl sarcosinate; 0.1% Triton X-100
- *MUG solution*: 1 mM MUG in GUS extraction buffer
- *MUG stop solution*: 0.2 M Na_2 CO_3
- *GUS staining buffer*: 100 mM sodium phosphate buffer pH 7.0 with 0.05% X-Glu dissolved in dimethyl formamide, in the presence of 0.1% sodium azide

2.2 Bacterial strains

Agrobacterium strain GV3101 (pPM6000) is a cured C58 nopaline strain containing the pTiAch5 derivative pPM6000, which is deleted in the T-DNA [10].

Agrobacterium strain GV3101 (pPM6000K) has been obtained by modification of the *virD2* gene of the Ti-plasmid. This strain lacks 70% of the coding sequence of the VirD2 protein, which is essential for T-DNA transfer. This strain was shown to be T-DNA transfer-defective [11].

2.3 *uidA* gene construct

pLRG [11] is a binary vector containing an improved *uidA* gene from pGUS23 [12, 13] in the T-DNA.

The *uidA* gene is under control of the 35S promoter from the cauliflower mosaic virus (CaMV), and the translational start site is derived from gene V of CaMV [12]. This modification of the translational start prevents the production of the GUS protein (β-glucuronidase) in the plant cells.

3   Methods

Protocol 1   Cocultivation of tobacco seedlings with *Agrobacterium*

1. 3 ml fresh overnight culture of *Agrobacterium* grown in YEB medium containing the antibiotics rifampicin (100 mg/ml) and gentamycin (40 mg/ml) is washed twice with 3 ml 10 mM mgSO and resuspended in 10 ml MS medium to a final $A_{600} = 0.6$.
2. 100 *Nicotiana tabacum* SR1 plantlets are added to the bacterial suspension. The plantlets are grown for one to two weeks on sterile, wet Whatman paper in growth chamber at 25 °C with 16 h light/day.
3. The mixture is exposed to reduce pressure (0.15 atm) in a sterile vacuum chamber for 5 min.
4. The seedlings are placed on MS plates (1% agar) and further cocultivated for three days in a growth chamber (25 °C 16 h light/day).
5. The plantlets are washed in sterile 10 mM $MgSO_4$ and blotted dry on sterile Whatman paper.
6. One part of the plantlets of each single cocultivation is analyzed with the fluorimetric MUG assay, another part with the histochemical GUS assay.

Protocol 2   Fluorimetric MUG assay

1. 30 plantlets are added to 400 ml of GUS extraction buffer and homogenized in a 1.5-ml Eppendorf tube.
2. The mixture is centrifuged for 10 min at 18,000 g at 4 °C. The supernatant is then considered as the protein extract.
3. The protein concentration of the extract is determined according to Bradford [14].
4. 10 μl of the extract is added to 200 μl of MUG solution in the first row of a multiwell plate (Dynatec Microfluor).
5. Immediately afterwards, 20 μl of the reaction mixture in the first row is transferred to 200 μl of stop solution added in the second row. This is the zero time-point for the enzymatic assay.
6. The plate is incubated at 37 °C.
7. Subsequently every hour, 20 μl of the reaction mixture in the first row is transferred to 200 μl of stop solution in the subsequent rows.
8. The fluorimetric signal of each sample is determined with a Titertek Fluoroskan II fluorimeter (excitation at 365 nm, emission at 455 nm).
9. The value of the enzymatic activity in each sample is calculated from the initial slope of the curve obtained by plotting the fluorimetric value against the time.
10. These values are normalized to the protein concentration of each extract, and the relative values (compared to the wild-type) of enzymatic activity of the different extracts are determined.

Protocol 3   Histochemical X-Glu assay

1. After cocultivation, 30 plantlets are added to 5 ml of GUS staining solution and soft vacuum (0.15 atm) is applied for 5 min in order to standardize the distribution of the substrate into the cells of different tissues in the plantlets. The reaction is allowed to proceed at 37 °C for 24 h.
2. The plantlets are washed in sterile water and bleached with ethanol.
3. The blue spots present on the plantlets are counted under a binocular.

4   Results and discussion

4.1   Interpretation of results

In order to assess the sensitivity of the assay, the transfer-proficient *A. tumefaciens* strain GV3101 (pPM6000, pLRG) was diluted in different ratios with the transfer-defective strain GV3101 (pPM6000K, pLRG), and the transfer efficiency was measured with the MUG assay (see Methods). A difference in the GUS activity of three orders of magnitude between the transfer-proficient strain and a 1-to-1000 dilution of this strain can be observed. Furthermore, even the value for the dilution 10^{-3} is clearly above the background (see Tab. 1), whereas further dilutions are showing a non-linear response, probably indicating the limit of detection of the assay. The values of the standard deviation given in Table 1 are very low. Thus, the values of the normalized enzymatic activity measured with the MUG assay can be considered reliable and precise.

Table 1   Detection of GUS activity upon T-DNA transfer to tobacco cells. The transfer proficient *Agrobacterium* strain was diluted in different rations with a transfer-defective strain.

Dilution proficient/defective	Enzymatic activity Normalized[1]	Blue spots Numbers[2]	normalized[3]
Proficient strain	1000	2100	1000
10^{-1}	120 ± 43	1530	730
10^{-2}	14 ± 4.0	631	300
10^{-3}	1.7 ± 0.25	267	130
10^{-4}	0.64 ± 0.20	20	10
10^{-5}	0.52 ± 0.10	1	0.5
defective strain	0.37 ± 0.025	0	0.0

[1]   MUG assay; activity of proficient strains normalized to 1000
[2]   X-Glu assay; number of blue spots in 30 plantlets
[3]   Number of spots of proficient strain normalized to 1000

In parallel with the MUG assay, a histochemical X-Glu assay was performed. In Table 1 we present the number of blue spots counted in one experiment after staining 30 plants per dilution. We found that an accurate counting of the number of spots is very difficult when the plantlets exhibit a large number of blue spots throughout the whole plant tissue. Spots of different color intensity also could be found, reflecting that either 1) transformation events occurred at different time periods through the incubation and as a consequence the quantity of GUS protein produced is variable from cell to cell or 2) the number of T-DNA per cell can vary. A comparison of the results obtained by the MUG assay with those obtained by the histochemical assay indicates that the histochemical assay underestimates the number of T-DNA transfer events when high concentrations of the transfer-proficient bacterial strain are involved. At low concentrations of the transfer-proficient strain (dilutions $1:10^3$ to $1:10^5$), where the number of blue spots can be determined more precisely, the histochemical X-Glu can be considered as reliable as the enzymatic MUG assay. Thus, the histochemical staining seems to be very adequate to detect very rare T-DNA transfer events.

4.2 Applications

Plant biotechnologists need to work on two key parameters to improve plant transformation efficiency: they have to identify plant tissues that can be both regenerated and susceptible to transformation. This can be very difficult, especially when starting to transform a new plant type or variety, because one would have very few indications whether the ideal transformation conditions are met. To help in that process, the *uidA* gene is frequently used as a means to select the appropriate transformation condition at a very early stage of the transformation. The *uidA* gene allows one to establish the best transformation protocol from the testing of different tissues, tissue culture conditions, and *Agrobacterium* strains in different combinations without having to wait for the complete and lengthy transformation process to be achieved. The histochemical protocol, using the X-Glu substrate, will also allow one to determine directly, *in planta*, which are the plant tissues showing the highest susceptibility to a given *Agrobacterium* strain.

5 Troubleshooting

In the assay described here, we measured the enzymatic activity of GUS, an enzyme produced upon transfer of T-DNA containing the gene for it. The assay is sensitive, with a low background activity; reproducible; and quantitative over a wide range of concentrations.

The high sensitivity of the assay is most probably due to the improved *uidA* gene construct used. The sequence surrounding the start codon is derived from gene V of CaMV [12]. This allows efficient translation, with negligible background activity stemming from *A. tumefaciens*. A similar construct was described earlier [15]. The remarkable sensitivity of the assay most probably also reflects the quality and competence of the tobacco seedlings to act as recipient for *Agrobacterium*-mediated DNA transfer.

We attribute the reproducibility of the assay to the fact that each measured value stems from a large number of seedlings. Thus, a statistically significant sampling is possible, avoiding plant to plant differences. The assay is quantitative over a wide range of concentrations of transfercompetent *Agrobacterium* cells. This applies to the fluorimetric assay, for which extracts of many seedlings guarantee standardized measurements. The background inherent to the system does not allow monitoring of T-DNA transfer efficiencies comparable to a 10,000-fold dilution of our standard strain. The histological staining is not accurate in the higher ranges of concentration (activity) because of many cotransfer events and confluence of blue spots. The first product of the reaction catalyzed by GUS is known to diffuse out from the plant cell of origin, but allows for a fairly precise measurement in the lower concentration (activity) ranges as a result of the non-existent background of a blue spot. A combination of the two methods of detection thus covers almost five orders of magnitude.

Both detection methods most probably assay transient expression of T-DNA not (yet) integrated in the plant genome [15, 16]. Thus, the gene activity of the T-DNA units that entered into the plant cell nucleus is measured, irrespective of integration. Therefore, problems in the interpretation of the results linked to position effect on gene expression can be avoided. GUS activity as measured by the fluorimetric assay is considered to be strictly dependent on the number of T-DNA units entering a plant cell nucleus. The measured values therefore are likely to be proportional to the number of T-DNA units entering the nuclei (no matter how many nuclei). In contrast, data from the histochemical staining are likely to be proportional to the number of nuclei that have received at least one expressing T-DNA unit.

The assay described above allows one to characterize bacterial parameters important for T-DNA transfer, such as the virulence gene activity of *cis*-acting sequences or other factors such as virulence gene inducers that contribute to the efficiency of T-DNA transfer.

In practice, the ability to monitor transfer of the *uidA* gene to plant cells has been of great help to plant biotechnologists since it has allowed efficient performance of extensive screens at an early stage of the transformation process (prior to regeneration). This has led to the optimization of bacterial (vectors, growth conditions) and plant tissue parameters (origin of tissue to be use, dividing stage, phytohormone balance) that conditioned T-DNA transformation efficiency of recalcitrant plant species.

Acknowledgments

I would like to warmly thank Prof. Barbara Hohn, in whose lab these methods were developed, as well as my former colleagues Dr. Jesus Escudero and Dr. Luca Rossi, who were major contributors.

References

1 Tinland B (1996) The integration of T-DNA into plant genomes. *Trends Plant Sci* 1: 178–184

2 Zhu J, Oger PM, Schrammeijer B, Hooykaas P et al. (2000) The bases of crown gall tumorigenesis. *J Bacteriol* 182: 3885–3895

3 Zupan J, Muth TR, Draper O, Zambryski P (2000) The transfer of DNA from *Agrobacterium* into plants: a feast of fundamental insights. *Plant J* 23: 11–28

4 Komari T, Hiei Y, Ishida Y et al. (1998) Advances in cereal gene transfer. *Curr Opin Plant Biol* 1: 161–165

5 Azpiroz-Leehan R, Feldman KA (1997) T-DNA insertion in *Arabidopsis*. *Trends Gen* 13: 152–156

6 Bouche N, Bouchez D (2001) Arabidopsis gene knock out: phenotype wanted. *Curr Opin Plant Biol* 4: 111–117

7 Murashige T, Skoog F (1962) A revised medium for rapid growth and bio-assays with tobacco tissue culture. *Physiol Plant* 15: 473–493

8 Sambrook J, Fritsch E, Maniatis T (eds) (1989) *Molecular cloning: a laboratory manual.* Cold Spring Harbor Press, Cold Spring Harbor, NY

9 Jefferson RA (1987) Assaying chimerci genes in plants: the GUS gene fusion system. *Plant Mol Biol Reporter* 5: 15–38

10 Bonnard G, Tinland B, Paulus F et al. (1989) Nucleotide sequence, evolutionary origin and biological role of a rear-ranged cytokinin gene isolated from a wide host range biotype III *Agrobacterium* strain. *Mol Gen Genet* 216: 428–438

11 Rossi L, Hohn B, Tinland B (1993) The VirD2 protein from *Agrobacterium tumefaciens* carries nuclear localization signals important for transfer of T-DNA to plants. *Mol Gen Genet* 239: 345–353

12 Schultze M, Hohn T, Jiricny J (1990) The reverse transcriptase gene from CaMV is translated separately from the capsid gene. *EMBO J* 9: 1177–1185

13 Puchta H, Hohn B (1991) A transient assay in plant cells reveals a positive correlation between extrachromosomal recombination rates and length of homologous overlap. *Nucleic Acids Res* 19: 2693–2700

14 Bradford M (1976) A rapid and sensitive method for the quantitation of microgram quantities of protein utilizing the principle of protein-dye binding. *Anal Bioch* 72: 248–254

15 Janssen BJ, Gardner RC (1989) Localised transient expression of GUS in leaf discs following cocultivation with *Agrobacterium*. *Plant Mol Biol* 14: 61–72

16 Machida Y, Shimoda A, Yamamoto-Toyoda Y et al. (1992) Molecular interactions between Agrobacterium and plant cells. In: EW Nester (ed): *Advances in molecular genetics of plant-microbe interactions*. Kluwer, Dordretch, 2: 85–96

10 Quorum Sensing: Approaches to Identify Signals and Signalling Genes in Gram-negative Bacteria

Simon Swift

Contents

Methods and Tools in Biosciences and Medicine
Prokaryotic Genomics, ed. by M. Blot
© 2003 Birkhäuser Verlag Basel/Switzerland

1   Introduction

Quorum sensing (QS) embodies the concept that bacteria do not simply grow as individual cells but rather act together as populations. QS is a generic mechanism for the regulation of diverse aspects of bacterial physiology including pathogenicity, symbiosis, biofilm formation, secondary metabolism, dissemination and dispersal, DNA transfer, and dormancy [1, 2]. QS relies upon small signal molecules to provide the bacterium-to-bacterium communication that effects the regulation of gene expression, a process that is an attractive target for a new generation of anti bacterial agents that have been termed quorum-sensing blockers (QSBs) [1, 2].

What is QS? In its original definition, QS is a mechanism for the sensing of the cell density of a population, such that as population size increases so does the concentration of a signal molecule, until a threshold is reached and perception of the signal induces a change in gene expression [3]. The process of QS can be investigated from the perspective of signal generation, the signal itself and the perception and response to the signal. Signal chemistry often comprises a modification or coupling of amino acid and/or lipid metabolism and undoubtedly reflects many niche-associated variables [4]. Signal perception may be intracellular, where sensing and regulatory activities reside in the same protein, or extracellular, where sensing and regulatory activities form a two-component perception and response unit. The methodology in this article will concentrate upon *N*-acyl homoserine lactone (acyl-HSL) signalling for which the term QS was originally coined. Here the predominant QS apparatus is the LuxI-type signal synthase and the LuxR-type intracellular sensor regulator. Figure 1 shows the basic theory of acyl-HSL mediated QS and the practical areas this article will address.

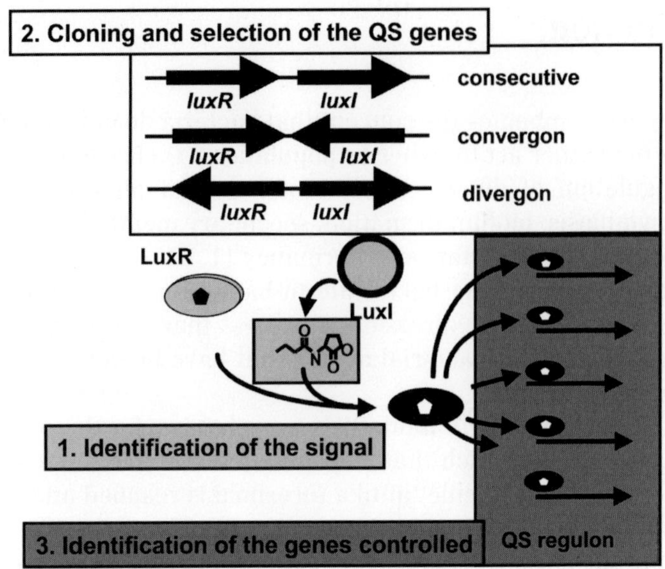

Figure 1   Acyl homoserine-mediated quorum sensing. Members of the *luxI* gene family encode the LuxI-type acyl-HSL synthase. Members of the *luxR* gene family encode the LuxR type acyl-HSL response regulator. The *luxI* and *luxR* homologues in a particular bacterium are often closely linked but have a gene organisation that can be consecutive, convergent, or divergent. The LuxI-type proteins are required for the biosynthesis of the acyl-HSL (C4-HSL is shown), which is a ligand for LuxR-type regulators (shown as a dimer). Gene activation of the QS regulon occurs as a result of the interaction of the LuxR-type regulator with its ligand, often *via* activation of transcription, as shown, although in some cases this ligand interaction is required for de-repression. In this article, methods for signal identification, QS gene cloning, and the identification of the QS regulon are described.

2   Materials

2.1   Chemicals and cell culture

Solvents (acetone, acetonitrile, dichloromethane, ethyl acetate, methanol) for extraction and purification of acyl-HSLs should be high-pressure liquid chromatography (HPLC) grade.

Anhydrous $MgSO_4$ (Sigma M7506) is used to remove water from solvent extracts.

For acyl-HSL fractionation by thin layer chromatography (TLC), use 20 cm × 20 cm reverse phase RP-18 F_{254s} (e. g., Merck 76012 2M) or normal phase silica gel 60 F_{254} (e. g., Merck 155622 2G) plates.

Synthetic acyl-HSLs, e. g., for use as standards, are available through Fluka and/or Sigma-Aldrich (product numbers in parentheses) if facilities for in-house biosynthesis are not available. Currently available are *N*-butanoyl-DL-homoserine lactone (09945); *N*-hexanoyl-DL-homoserine lactone (C6-HSL; 09926); *N*-

heptanoyl-DL-homoserine lactone (10939); *N*-octanoyl-DL-homoserine lactone (10940); *N*-decanoyl-DL-homoserine lactone (17248); *N*-dodecanoyl-DL-homo-serine lactone (17247); *N*-tetradecanoyl-DL-homoserine lactone (10037); *N*-3-oxohexanoyl-DL-homoserine lactone (K3255), also known as *N*-(β-ketocaproyl)-DL-homoserine lactone and *N*-3-oxohexanoyl-L-homoserine lactone, also known as *N*-(β-ketocaproyl)-L-homoserine lactone (K3007). The L-form is presumed to be the active form in nature.

In the absence of specific equipment for imaging bioluminescence, photo-graphic film such as that used for imaging enhanced chemiluminescence (ECL) blots (e. g., Amersham/Pharmacia Hyperfilm ECL) is suitable.

2.2 Equipment

For signal extraction and purification, a fume hood, glass-separating funnels (up to 2 l capacity), a rotary evaporator, and an inert gas (e. g., N_2) airline are required. For HPLC use a reverse-phase preparative column (e. g., C8, such as Kromasil KR100–5C8, 250 mm × 8 mm).

For the detection of low-level bioluminescence in macroscopic and micro-scopic situations, photon imaging cameras are recommended (e. g., Berthold Technologies Night Owl). For the assay of bioluminescence, use single tube and/ or microplate format luminometers. A large range of products is available from a number of suppliers including Anthos-Labtec instruments, Berthold Technol-ogies, Perkin Elmer Life Sciences, and Turner Designs.

For the introduction of gene libraries into *Escherichia coli*, an electroporator (BioRad, set at 2.5 kV, 25 µF, 200 Ω; 0. 2 cm electroporation cuvette) is recommended.

2.3 Solutions, reagents and buffers

Complex media for the growth of bacteria include L-broth (10 g l⁻¹ tryptone, 5 g l⁻¹ yeast extract, 5 g l⁻¹ sodium chloride) and TY broth (8 g l⁻¹ tryptone, 5 g l⁻¹ yeast extract, 2.5 g l⁻¹ sodium chloride). Agar is added at 1.2% weight:volume (w:v) or, for soft agar, 0.65% w:v. For blue/white screening of recombinants, isopropyl-1-thio-(-D-galactopyranoside) (IPTG) and 5-bromo-4-chloro-3-indo-lyl-(-D-galactoside) (X-Gal) are added to a final concentration of 0.1 mM and 40 µg ml⁻¹, respectively. Specified antibiotics are used at 50 mg l⁻¹ (ampicillin), 10 mg l⁻¹ (tetracycline), or 30 mg l⁻¹ (chloramphenicol, kanamycin).

3   Methods

The study of quorum sensing essentially covers three basic goals: 1) identification of the signal(s), 2) cloning of the genes involved in perception and response, and 3) determining the genes that are controlled. This section details strategies (see Fig. 2) and protocols that can be used to achieve these goals, the fulfilment of which is central to the search for QSBs.

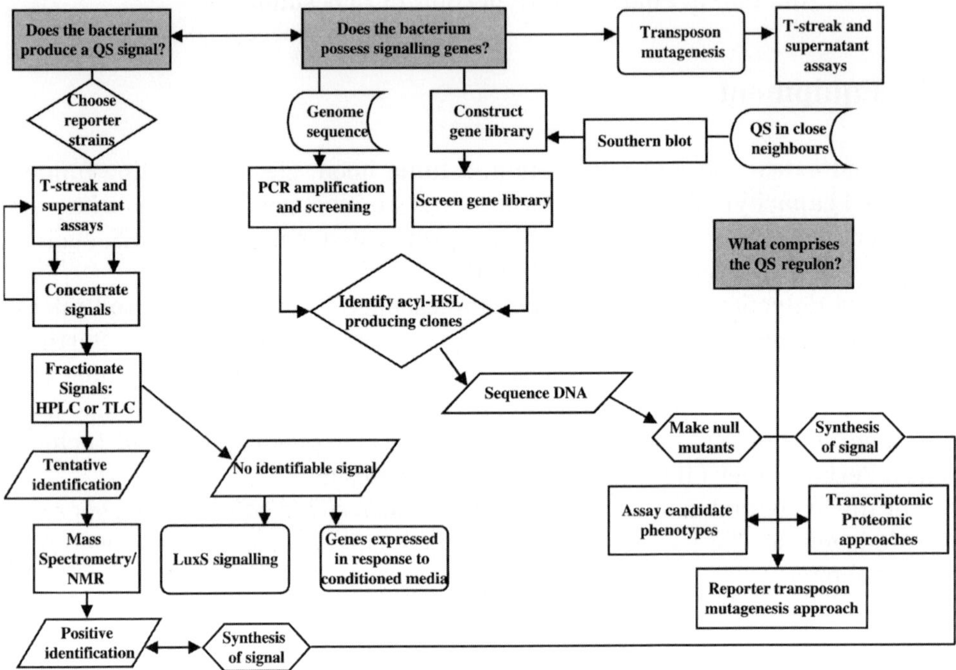

Figure 2   Strategies for the elucidation of the elements of quorum sensing. The questions asked in QS research have a grey background. Experimental processes are in rectangles, alternative strategies are in rounded rectangles, decisions are in diamonds, data obtained is in trapezoids, and preparations that may be required are in hexagons.

3.1   Identification of the signal

In cases where signalling in a specific environment or under specific conditions is being investigated, an important early step is to detect the signal in those conditions.

Protocol 1　Reporter strains

The biosensor used will detect the signals it is most sensitive to and may not give much, or any, response to what may turn out to be the major signal. A range of reporter strains exists to test whether a strain produces an acyl-HSL signal (Tab. 1)). Table 2 gives an indication of the suitability of different biosensors for a given acyl-HSL.

Chromobacterium violaceum CVO26 is a versatile and easy-to-use reporter strain [5]. The plasmids pSB401, pSB536, and pSB1075 use promoter fusions to the *luxCDABE* operon of *Photorhabdus luminescens*. Each is normally transformed into *Escherichia coli* for use as a biosensor. The group provides detection of a wide coverage of chain-length variability [6, 7].

Protocol 2　T-streaks and cross streaks

The test bacterium is streaked up to, or across, a streak of the reporter strain on an agar plate [7]. Activation of the reporter, e.g., after overnight incubation, indicates the production of a signal. Allow sufficient spacing, as highly active test strains may obscure the results from adjacent test strains.

Protocol 3　Supernatant assays

QS signal molecules accumulate in the culture supernatant. Spent culture supernatant can be obtained after centrifugation of a culture to remove the majority of the bacterial cells. For microplate well assays [6]:

1. Place 100 µl of test solution into the microplate well and perform any doubling dilutions into fresh growth medium as necessary. Endpoint dilutions are useful, as supernatants may contain compounds that inhibit the growth of the reporter strain. Dilutions of a synthetic standard or positive control supernatant are advised.
2. Place 100 µl of biosensor into each well and mix. For optimal results, it is important to prepare the biosensor prior to addition. For the pSB series, a 1:10 dilution in fresh medium is advised. Similar results are obtained with a larger dilution (1:100) followed by growth to mid-exponential phase (A_{600nm} = 0.5). For *C. violaceum* CVO26, a significantly better signal is obtained by diluting an overnight culture of the sensor 1:50 (v:v) into media containing agar at 1%.
3. Incubate at the advised temperature and visualise. The *lux* sensors are best suited to this type of analysis and will often give a quantitative signal after 30 min, which is optimal after 4–6 h. To visualise bioluminescence, a number of microplate readers and photon imaging cameras are now available. Photographic film can also be used, where 30 s is often enough for bright sensors, e.g., those containing pSB401, whereas 5 or more minutes may be necessary for the less active sensors, e.g., containing pSB1075.

Table 1 Useful bacterial strains and plasmids

Strain/Plasmid	Description	Uses	Reference
Bacterial strains			
Agrobacterium tumefaciens NT1	*A. tumefaciens* NT1 (no Ti plasmid)	Does not produce acyl-HSLs, host strain for sensor plasmid pDCI41E33	[10]
Chromobacterium violaceum CVO26	Double mini-Tn5 mutant derived from *C. violaceum* ATCC31532; Km^R, Hg^R	Acyl-HSL sensor, producing purple pigment violacein in the presence of activating (C4-C8) signal. >C8 may be detected using the reverse assay. A sensitive detector of C6-HSL. Incubate at 30 ºC.	[5]
Escherichia coli DH5α	F-, φ80*lacZ*, Δ*M15*, Δ(*lacYZA-argF*)$_{U169}$, *deoR, recA1, endA1, phoA, hsdR17*(r_{K^-}, m_{K^+}), *supE44*, λ⁻, *thi-1, gyrA96, relA1, luxS*	General purpose cloning strain, particularly useful for *luxS* studies	Invitrogen
Escherichia coli JM109	*mcrA, recA1, endA1, thi-1, hsdR17* (r_{K^-}, m_{K^+}), *supE44, gyrA96, relA1*, Δ(*lac-proAB*) [F' *traD36, proAB, lacI*^{q}, Δ*M15*]	General purpose cloning strain	Stratagene
Escherichia coli S17/1	*thi, pro, hsdR17*(r_{K^-}, m_{K^+}), *recA*, RP4 2-Tc::Mu-Km::Tn7, Tp^R Sm^R	Conjugal transfer of broad host range biosensor plasmids. λ pir lysogens will maintain RP4 suicide plasmids used in gene knockout protocols.	[6]
Serratia sp. ATCC 39006 LIS	*smaI*::mini-Tn5Sm/Sp, Sp^R	Acyl-HSL sensor, producing red pigment prodigiosin in the presence of activating signal. A sensitive detector of C4-HSL. Incubate at 30 ºC.	[24]
Pseudomonas aureofaciens 30–84I	*P. aureofaciens* 30–80 phzI::kan, Km^R	Acyl-HSL sensor, producing diffusible orange pigment phenazine in the presence of activating signals with acyl chains C4-C8. Incubate at 28 ºC.	[25]
Rhizobium leguminosarum A34	*R. leguminosarum* 8401 (cured of symbiotic plasmid) carrying pRL1JI (symbiotic plasmid of R. *legumino-sarum* bv *vicia*	Sensor for 3-OH-C14:1–HSL, this strain will not grow in the presence of this signal also known as bacteriocin small	[15]
Vibrio harveyi BB130	Environmental isolate	Positive control for LuxS and 3-OH-C4-HSL signals	[20]

Strain/Plasmid	Description	Uses	Reference
Vibrio harveyi BB170	luxN::Tn*5*	Sensor for LuxS signal, producing bioluminescence early in growth in response to signal. The sensor is unable to perceive the presence of 3-OH-C4-HSL, and so bioluminescence is induced only by the LuxS signal. The sensor is able to produce the LuxS signal itself, and so the assay depends upon the detection of premature induction of bioluminescence in comparison to a negative control.	[20]
Vibrio harveyi BB886	luxQ::Tn*5*	Sensor for 3-OH-C4-HSL signal, producing bioluminescence early in growth in response to signal. The sensor is unable to perceive the presence of the LuxS signal, and so bioluminescence is induced only by the acyl-HSL signal. The sensor is able to produce the acyl-HSL signal itself, and so the assay depends upon the detection of premature induction of bioluminescence in comparison to a negative control.	[20]
Plasmids			
pBluescript SK+ II	High-copy, multi-cloning site vector, *lacZ, oriC*, Ap^R	Cloning vector with blue/white selection. Compatible with ori p15a and oriV plasmids for acyl-HSL synthase cloning.	Stratagene
pDCI41E33	IncQ, Km^R, *traR, traG::lacZ*	Sensitive acyl-HSL sensor, when used to transform A. tumefaciens NT1, producing a blue pigment in the presence of X-gal as a result of β-galactosidase activity.	[10]
pHG327	pBR322 copy multicloning site vector, *lacZ, oriC*, Ap^R	Cloning vector with blue/white selection. Compatible with *ori* p15a and *oriV* plasmids for acyl-HSL synthase cloning. Recommended for cases where acyl-HSL synthase genes appear to be unstable at high copy number.	[26]
pJBA89	*luxR luxI'* (*Vibrio fischeri* ATCC7744)::RBSII(pQE70)::*gfp*(ASV)-T$_0$-T$_1$ *oriC*, Ap^R	Acyl-HSL biosensor with optimised ribosome binding site producing green fluorescence in the presence of acyl-HSLs. GFP (ASV) has a half-life of approximately 45 min giving low background fluorescence after 20 h incubation at 30 ºC.	[23]

Strain/Plasmid	Description	Uses	Reference
pJBA132	*luxR luxI'* (*Vibrio fischeri* ATCC7744)::RBSII(pQE70)::*gfp*(ASV)-T_0-T_1, p15a *ori*, pVS1 *ori*	Broad host range acyl-HSL biosensor producing green fluorescence in the presence of acyl-HSLs. GFP (ASV) has a half-life of approximately 45 min giving low background fluorescence after 20 h incubation at 30 ºC.	[23]
pRK415	*oriT, oriV, lacZ*, Tc^R	Broad host range cloning vector, blue/white selection. Compatible with *oriC* and *ori* p15a plasmids.	[27]
pSB401	*luxR luxI'* (*Vibrio fischeri* ATC-C7744)::*luxCDABE* (*Photorhabdus luminescens* ATCC29999), *ori* p15a, Tc^R.	Acyl-HSL biosensor, producing strong bioluminescence in the presence of 3-oxo-C6-HSL and related signals with acyl chains C6-C8. Compatible with *oriC* and *oriV* plasmids (note pRK415 and pSB401 share Tc^R) for acyl-HSL synthase cloning. Incubate at 30 ºC.	[6]
pSB403	*luxR luxI'* (*Vibrio fischeri* ATC-C7744)::*luxCDABE* (*Photorhabdus luminescens* ATCC 29999), *oriV, oriT*, Tc^R.	Broad host range acyl-HSL biosensor, producing strong bioluminescence in the presence of 3-oxo-C6-HSL and related signals with acyl chains C6-C8. Compatible with *oriC* and p15a *ori* plasmids for acyl-HSL synthase cloning. May be transferred by conjugation to a broad range of species for intracellular sensing of acyl-HSL production. Incubate at 30 ºC.	[6]
pSB536	*ahyR ahyI'* (*Aeromonas hydrophila* AH-1N)::*luxCDABE* (*Photorhabdus luminescens* ATCC 29999), *oriC*, Ap^R	Acyl-HSL biosensor, producing strong bioluminescence in the presence of C4-HSL and C6-HSL. Compatible with p15a *ori* and *oriV* plasmids for acyl-HSL synthase cloning.	[7]
pSB1075	*lasR lasI'* (*Pseudomonas aeruginosa* PAO1)::*luxCDABE* (*Photorhabdus luminescens* ATCC 29999), *oriC*, Ap^R	Acyl-HSL biosensor, producing bioluminescence in the presence of signals with acyl chains of C10 and above. Compatible with p15a *ori* and *oriV* plasmids for acyl-HSL synthase cloning.	[6]
pSU18 and series	Multicloning site vectors, *lacZ, ori* p15a, Cm^R	Cloning vector with blue/white selection. Compatible with *oriC* and *oriV* plasmids	[28]

For agar plate well assays [5]:

1. Prepare soft agar (e. g., LB, TY with 0.65% agar) and cool to 55 °C.
2. Add biosensor cells at 1:30 (v:v) dilution (i. e., 100 µl per 3 ml of agar) with appropriate antibiotics and pour 3 ml onto a plate of the same agar.
3. Allow the agar to set and cut a hole in the agar, e. g., with the "wrong end" of a sterile 1 ml tip or a sterile Pasteur pipette.
4. Mix the test sample with more soft agar and pipette into the hole, allow the agar to set, and incubate. Six holes may be spaced in a ring about 1–2 cm in from the edge of the agar plate for samples under test, with a central hole used for a positive control.
5. Visualise *lux* reporters using a photon imaging camera or photographic film.

Rhizobium growth inhibition assay: Bioassay plates are prepared by overlaying TY agar plates with 3 ml TY soft agar containing 100 µl of a stationary-phase culture of *R. leguminosarum* A34 (Tab. 2). Synthetic 3-OH-C14:1-HSL (1 µg ml^{-1}) or active *Rhizobium* supernatants may be used as a positive control. Plates are incubated at 28 °C for 48 h and examined for zones of growth-inhibition activity around the wells.

Table 2   Useful biosensors for common acyl-HSLs

Acyl-HSL (abbreviation)	Biosensor. Black fill indicates a biosensor suitable for detection of the signal, Grey fill indicates that the senor will detect the signal at high concentration.							
	AT	CV	CVR	pSB401	pSB536	pSB1075	RL A34	BB886

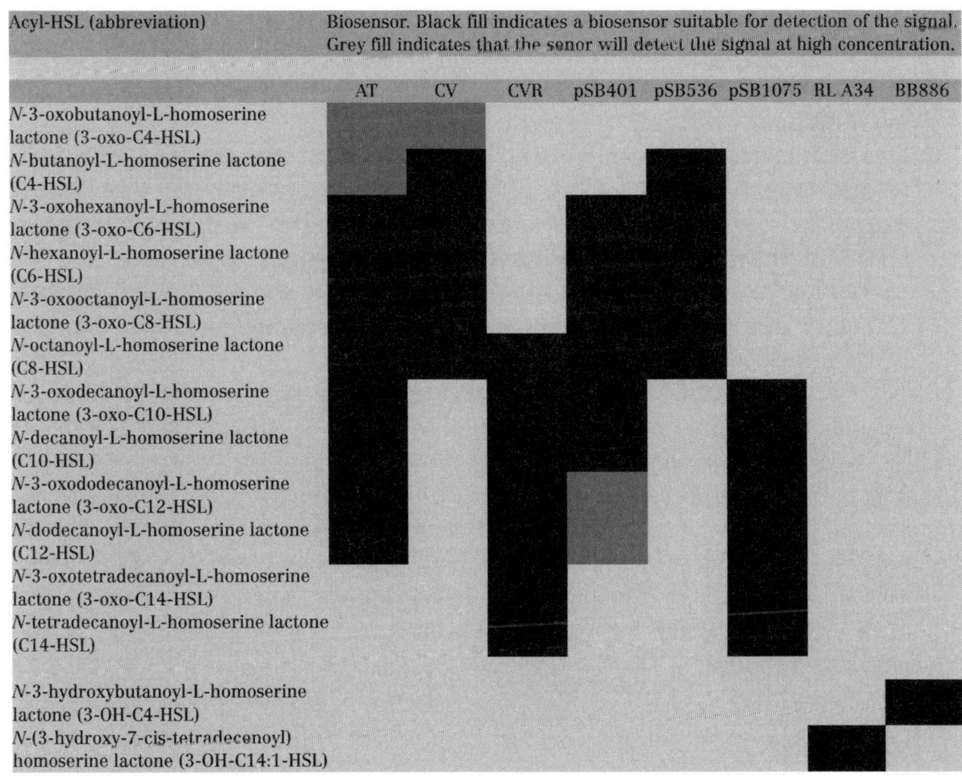

Acyl-HSL (abbreviation)
N-3-oxobutanoyl-L-homoserine lactone (3-oxo-C4-HSL)
N-butanoyl-L-homoserine lactone (C4-HSL)
N-3-oxohexanoyl-L-homoserine lactone (3-oxo-C6-HSL)
N-hexanoyl-L-homoserine lactone (C6-HSL)
N-3-oxooctanoyl-L-homoserine lactone (3-oxo-C8-HSL)
N-octanoyl-L-homoserine lactone (C8-HSL)
N-3-oxodecanoyl-L-homoserine lactone (3-oxo-C10-HSL)
N-decanoyl-L-homoserine lactone (C10-HSL)
N-3-oxododecanoyl-L-homoserine lactone (3-oxo-C12-HSL)
N-dodecanoyl-L-homoserine lactone (C12-HSL)
N-3-oxotetradecanoyl-L-homoserine lactone (3-oxo-C14-HSL)
N-tetradecanoyl-L-homoserine lactone (C14-HSL)
N-3-hydroxybutanoyl-L-homoserine lactone (3-OH-C4-HSL)
N-(3-hydroxy-7-cis-tetradecenoyl) homoserine lactone (3-OH-C14:1-HSL)

Biosensors are *A. tumefaciens traG::lacZ* (AT), *C. violaceum* CV026 (CV), *C. violaceum* CV026 reverse assay (CVR), *E. coli*/pSB401 (pSB401), *E. coli*/pSB536 (pSB536*), E. coli*/pSB1075 (pSB1075), *R. leguminosarum* A34 (RLA34) and *V. harveyi* BB886 (BB886).

Chromobacterium violaceum CV026 reverse assay detects long-chain acyl-HSLs through their antagonism of activating signals. Include C6-HSL at 5 µM in the test agar (this protocol, step 2) and look for zones of no pigmentation in the overlay. Ensure that positive reverse assay results are not scored for compounds that inhibit sensor growth.

Protocol 4   Solvent extraction of signal molecules

Organic solvents may be used to extract and concentrate signal molecules from supernatants (and also from bacterial cells or from biofilms and other natural sources) [8, 9]:

1. Large-scale extractions (typically 1 l or greater volume of supernatant) are normally performed to provide sufficient signal for mass spectrometry. Supernatants are vigorously shaken with an organic solvent in a separating funnel. Vapourisation of the solvent will increase gas pressure within the separating funnel; this should be periodically released during the extraction for safety. Dichloromethane has a greater density than water and partitions as the lower phase. Each aliquot of media is extracted three times with 0.7 volumes of the solvent. In some cases, surfactants in the media necessitate a long settling time and/or centrifugation to separate the phases. The growth of the bacterium in a defined minimal medium, rather than a complex medium, may reduce this emulsification.

2. Small-scale extractions (typically from a 5–50 ml culture) are often performed to qualitatively assess the QS signals present and may use a single extraction. Small separating funnels may be used with dichloromethane. For extractions in glass tubes, ethyl acetate is less dense than water and will partition as the upper phase from where it may be removed with a pipette. The compatibility of solvents and any plasticware should be checked if glass is not to be used.

3. The recovered solvent should be dried of any water using anhydrous magnesium sulphate, filtered through Whatmann No 1 paper to remove the hydrated and residual anhydrous salt, and then dried by rotary evaporation. The residue can be reconstituted in either the extraction solvent or the organic solvent used in HPLC applications, e. g., acetonitrile or methanol. Reconstitution in a new solvent may leave a residue of insoluble, non-signal, material. It is important to thoroughly clean any glassware used in the extraction and concentration process to avoid contamination of the next extraction.

4. Assay of solvent extracts requires dilution of the solvent below levels that are toxic to the reporter strain. This may be achieved by performing dilutions into the aqueous test medium (where a solvent-only control should be included) or by performing dilutions in the solvent, evaporating the solvent, and adding the reporter strain in aqueous medium to reconstitute the signal.

Protocol 5   TLC assay

Reverse-phase TLC (Si-C_{18} plates) with a mobile phase of 60% (v:v) methanol in water is well suited to the fractionation of shorter chain acyl-HSLs [5, 7, 8, 10]. Normal phase TLC is recommended for longer chain acyl HSLs using a mobile phase of 55% hexane/45% acetone.

1. Mark the plate gently using a soft pencil to indicate where each sample and marker is to be spotted. Place the origin 3 cm from the bottom of the plate and the samples at least 2 cm apart.
2. Spot 1 to 10 µl of the extracted supernatant. Take care to make a small, even spot using, e. g., a hairdryer to dry aliquots of 0.5 to 1 µl during loading. For positive controls, a combined 3-oxo series (3-oxo-C4-HSL, 3-oxo-C6-HSL, 3-oxo-C8-HSL) and a combined unsubstituted series (C4-HSL, C6-HSL, C8-HSL) is recommended. Table 3 gives an indication of the amount of standard to be loaded for some of the biosensors.

Table 3   Acyl-HSLs as standards in TLC overlays
The figures indicate the concentration of acyl-HSL required to give a detectable signal for the biosensor indicated when 10 µl is loaded to a TLC plate. Note the figures given for pSB1075 are for normal phase TLC plates. ND designates an acyl-HSL that cannot be detected by the given biosensor at amounts below 10 µl of a 1 mg ml^{-1} solution.

Acyl-HSL: add 10 µl	CV026	pSB401	pSB1075
	µg ml^{-1}	µg ml^{-1}	µg ml^{-1}
C4-HSL	100	ND	ND
C6-HSL	10	10	ND
C8-HSL	1000	10	1000
C10-HSL	1000	100	100
C12-HSL	1000	100	10
3-oxo-C4-HSL	1000	1000	ND
3-oxo-C6-HSL	100	1	ND
3-oxo-C8-HSL	100	1	1000
3-oxo-C10-HSL	1000	10	10
3-oxo-C12-HSL	1000	100	1
3-oxo-C14-HSL	1000	100	10

3. Equilibrate the tank with the mobile phase, place pieces of Whatmann No. 1 on the inner side of the tank, and allow capillary action to wet them. The mobile phase should be about 1 to 2 cm in depth, so as not to come over the origin on the loaded TLC plate.

4. Place the loaded plate into the tank, two plates can be run together. The reverse-phase plates normally take 2 to 4 h to develop, after which the plates are removed, the solvent front marked for R_f calculations, and any solvent on the plate is evaporated off.

5. The TLC plate is overlaid with the biosensor (5 ml of an overnight culture in 100 ml of 0.5%-1% L-agar for *E. coli* containing pSB series plasmids and/or *C. violaceum* CV026). This requires either construction of a pouring device or simply taping the edges of the plate. The plate should be placed in a clean container for incubation.

6. Incubate overnight for *C. violaceum* CVO26 and other pigment-producing sensors; the *lux* sensors may be visualised after a shorter period of time.

7. The use of aluminium backed TLC plates facilitates preparative TLC. Here the plate is loaded as a line with markers in an end lane. After the TLC plate is developed, the plate can be cut to remove the markers and about 1 cm of the test for signal visualisation with a biosensor. Silica at the same position as the unknown spot can then be removed and any acyl-HSLs eluted from the silica using acetone, which can be dried and the residue reconstituted for, e. g., mass spectrometry. For more information regarding HPLC fractionation, mass spectrometry, NMR spectroscopy, and other detection assays for acyl-HSLs see [11–14].

3.2 Gene cloning

If the genome sequence of the organism of interest is available, gene amplification by PCR can be used to amplify the genes for cloning. If the genome sequence is not available, a gene library is required to screen for signal producers [7, 8, 15].

Protocol 6 Construction and screening of a gene library

1. Prepare genomic DNA libraries for the test bacterium using restriction endonucleases that cut the DNA to leave a good distribution of fragments in the 1 to 8 kb range and ligate into a suitable vector. Homologues of *luxI* are sometimes not stable at the high copy number provided by many of the commercially available cloning vectors. The use of lower copy vectors, e. g. pHG327, pSU series, should be considered, although if the copy number is too low, insufficient signal may be made.

2. Use the ligation to transform either *E. coli* or a reporter strain, e. g., *E. coli* (pSB401). If the cloning vector used has the facility for insertional inactivation of *lacα*, blue/white screening on media containing IPTG and X-gal can be performed to check the insert frequency. In some cases cloning into E.

coli is not recommended and other species should be considered, e. g., A. tumefaciens C58.00 [15].

3. Screening a library. Replica pick each recombinant (white colonies if blue/ white selection is possible) and grow overnight, allowing time for the clone to produce signal. T-streak or overlay with the sensor of choice and incubate overnight again before visualisation. If the library is made in a reporter strain, e. g., *E. coli* (pSB401), each positive clone will give rise to a bioluminescent colony. If access to a photon video camera is available, these colonies can be easily visualised. It is important to have a plating density of 100 to 500 colonies, otherwise, clone selection will be difficult. Alternatively, patch out recombinants and screen for positive clones using photographic film or microplate luminometry. The library should be grown at 30 ºC, as the LuxR activator protein does not work well at 37 ºC.

4. Plasmid DNAs should be isolated form the positive clone and re-transformed into *E. coli*, with selection only for the antibiotic resistance of the cloning vector, to purify the clone form the sensor plasmid. This strain will be able to activate the sensor plasmid in a T-streak but will be sensitive to the antibiotic of the sensor plasmid and will be dark.

5. DNA sequencing of the insert DNA will identify the *luxI* homolog. The acyl-HSL synthase may be a member of the LuxM family, the HdtS family, or a new family.

3.3   The QS regulon

The first reports of QS control were deduced after the investigation of regulatory mutants for given phenotypic traits. Subsequently, many investigations of QS have focussed first on the signalling system and then gone on to look at what is controlled. Even if an investigation has taken the first route it is likely that the phenotypic trait under study is but one of a number of genes regulated as part of a high cell density regulon and that the identification of co-regulated genes may be important.

Protocol 7   Signalling gene knock outs

1. It is necessary to make mutants in the *luxI* and *luxR* homologues. There are many strategies and techniques available, which are not described here [16–18]. It is essential to consider the applications to which the mutant may be put before designing the mutation strategy. Marking the strain, e. g., with an antibiotic resistance or GFP, may enhance future work on comparative localisation of bacteria in biofilms or during pathogenicity and in comparative colonisation and survival studies. Marker exchange mutagenesis will allow antibiotic selection of the mutant in these situations. Deletion of at least a portion of the gene will guard against reversion, especially in virulence studies.

2. The mutant should be compared to the parent for signal production, as many bacteria possess a second signalling system.

Protocol 8   What is the phenotype of the null mutants?

A number of random and directed strategies can be employed [7, 16].

1. QS controls traits that are activated at a high cell density: Qualitative assays of these traits (e. g., on agar plates detecting secreted exoenzymes, pigments) followed by quantitative assays throughout growth may identify regulated activities.

2. Transposons have been used to identify regulated characteristics. A promoterless reporter gene on a transposon (e. g., mini-Tn5 *luxAB*) is used to mutagenise the mutant strain. The mutant library is replica plated onto media ± synthetic signal. Activated genes will be ON (i. e. bioluminescent if *luxAB*) in the + (i. e., with exogenous signal) situation and OFF in the − situation [19].

3. The addition of an exogenous signal can be used as an alternative to, or forerunner of, mutant studies. The signal identified (see section 3.1) is simply added at the start of growth and phenotypic traits are assayed through growth.

4. Proteomic and transcriptomic approaches: Alterations in transcription and protein expression that are regulated by QS may be identified by mutant analysis and signal addition experiments.

4   Troubleshooting

- *A T-streak or supernatant assay fails to identify the presence of a signal:* The conditions for signal production may not be optimal, and changing the growth media or temperature may help. Some bacteria may signal only in the conditions encountered in nature. Replication of these conditions in the laboratory may identify signal production. Incubating the test culture longer may allow more of the signal to accumulate. Conversely, overlong incubation of the test culture may result in the degradation of any signal present. If the signal is present at only a low level, it may be concentrated to detectable levels after solvent extraction from culture supernatant.

- *Changing the growth conditions and concentrating the supernatant fails to identify a signal:* The signal may still be present, but other compounds present in the culture supernatant prevent detection. HPLC or TLC fractionation of the concentrated extract is likely to separate the signal from any inhibitory compound, allowing detecting by TLC overlay or assay of HPLC fractions.

- *Fractionation of the concentrated supernatant fails to identify a signal:* Search the genome sequence and/or attempt to clone an acyl-HSL synthase for completeness. If acyl-HSL signalling has been ruled out, investigations should examine the possibility of other signalling molecules. LuxS signalling is present in many species of both gram-negative and gram-positive bacteria.

Parallel investigation of LuxS and acyl-HSL signalling also may prove profitable. The LuxS signal can be detected by using a *Vibrio harveyi* biosensor strain (see Tab. 1) [20].

An alternative approach is to assume that cell density-dependent regulation of gene expression is occurring, but have no prejudice for signal chemistry. Reporter transposons can then be used to identify genes expressed at high cell density, and any involvement of signalling can then be investigated through the analysis of spent media and cross-feeding experiments [21].

- *The T-streak is positive but there is little or no signal in the concentrated organic extract:* Broth conditions may not favour acyl-HSL production; therefore, the agar plate conditions should be mimicked as closely as possible. The signal may have been made and then degraded during incubation of the broth culture. To overcome these problems, small-scale solvent extractions of multiple cultures can test different growth conditions and the time points in a growth curve.

- *Tentative identification of the signal is not possible or positive identification based on the tentative identification is not possible:* If there is no tentative match, or if the mass spectra of the unknown and its tentative match are different, it is likely that the signal is a "novel" molecule. In this case, structure determination using NMR is required.

- *There is no* luxI *homolog in the genome sequence or the DNA sequence of an active clone:* The LuxM and HdtS acyl-HSL synthases also exist [8, 17]

- *The mutant and parental strains show no obvious phenotypic differences:* QS is most commonly associated with the regulation of traits associated with growth at a high-population cell density. Candidates for QS regulation should be assayed quantitatively and throughout growth, if possible, to ensure that activation *via* other mechanisms does not mask any effects of QS. Consider the assay conditions also; is the QS signal being made by the parental strain?

- *Reporter transposons fail to identify activated genes:* Multiple regulatory inputs may activate gene expression over the course of an incubation to levels that are similar in QS mutant and parent. QS effects may be seen as temporal variations in the activation of this gene expression and, where possible, the activity of the reporter gene should be assayed throughout growth.

- *The addition of exogenous signal has no observable effect:* Prevailing conditions may repress gene expression, and QS may activate only once that repression is relieved.

- *QS mutants with identifiable regulated characteristics revert to the parental phenotype:* QS is a regulatory mechanism and the structural genes for the various traits remain. Original mutants should be stored at −80 °C to guard against any reversion events.

5   Applications

5.1   When all goes according to plan

In choosing applications to demonstrate the methodologies and strategies described, *Aeromonas hydrophila* presents itself as a case where the progress went according to plan [7, 18]. The bacteria produce a positive T-streak on agar, and concentrated extracts can be shown by TLC to contain C4-HSL and C6-HSL. Mass spectrometry of the HPLC-purified material confirmed this. Chromosomal inactivation of the *ahyI* gene resulted in a substantial reduction in extracellular protease activity, which was recovered by the addition of exogenous signal. Protease was originally hypothesised as a regulated characteristic because of its known stationary phase induction. This effect upon protease activity was observable on agar plates, in quantitative assays of supernatant proteins, and by polyacrylamide gel electrophoresis analysis of secreted proteins.

5.2   Where is the signal, where is the gene?

T-streaks of *Pseudomonas fluorescens* F113 fail to activate biosensors for either short-chain or long-chain acyl-HSLs. Nevertheless, solvent extraction and HPLC fractionation of spent supernatants revealed the presence of three signal molecules, and the gene responsible for directing the synthesis of 3-OH-C14:1-HSL, C10-HSL, and C6-HSL was cloned by complementation of *E. coli* JM109 (pSB401). Analysis of the DNA sequence did not reveal the presence of a *luxI* homolog, but a potentially new type of acyl-HSL synthase was discovered. An open reading frame with homology to HdtS, a lysophosphatidic acid acyl-transferase, was identified in the clone; disruption of this gene on the plasmid removed the ability of the *E. coli* host to make a signal. To demonstrate that *P. fluorescens* HdtS possessed acyl-HSL synthase activity, the protein was pro-duced *via in vitro* transcription and translation and assayed for activity *in vitro* using defined substrates [8].

5.3   The signal molecule is not an acyl-HSL

Acyl-HSLs are not the only molecules that are extractable with an organic solvent and able to activate biosensors. In cell-free *Pseudomonas aeruginosa* culture supernatants, two such compounds were identified. Mass spectrometry and NMR spectroscopy revealed that these compounds were the diketopiper-azines (DKPs). Activation of bioluminescence by these DKPs, from the LuxR-based biosenor encoded upon pSB401 in *E. coli,* is dose-responsive, although this is at much higher concentrations than the natural activator 3-oxo-C6-HSL. Furthermore, competition studies showed that the DKPs antagonize the 3-oxo-

C6-HSL-mediated induction of bioluminescence, suggesting that the DKPs may compete for the same LuxR-binding site. Supernatants from a range of gram-negative bacterial species contain various cyclic dipetides. Importantly, these molecules are also present in fresh culture media, e. g., L-broth, that are rich in peptides, showing the value of a fresh-media negative control [13].

5.4 Transposon mutagenesis to identify phenotypic characteristics

In *Serratia liquefaciens* QS is required to activate surface translocation by swarming motility, with mutations in the *swrI* gene conferring a swarm-negative phenotype unless the acyl-HSL signal C4-HSL is exogenously added. Random mutagenesis of the *swrI* null mutant with a mini-Tn*5* derived transposon carrying the promoterless *luxAB* genes isolated 19 double mutants in which bioluminescence was induced in the presence of 200 nM C4-HSL. One of these mutants was unable to swarm in the presence of exogenous signal. Further analysis showed that the transposon insert is in *swrA*, a gene encoding a putative peptide synthase for the surfactant serrawettin W2 [19].

6   Remarks and conclusions

Application of the methodologies above has led to a golden age of discovery for acyl-HSL signalling. The availability of genome sequences, powerful analysis programs, and protocols for transcriptomic and proteomic analyses will short-cut many of the procedures described, but the experiments will still need to be performed. More frequently, the real-world relevance of signalling is being addressed, with signal production assayed in real situations. Often, this requires analysis of much smaller amounts of signal and the development and validation of more sensitive methods of signal detection. Moreover, the ability to demonstrate signalling at the level of the individual cell within a population, e. g., *in vivo*, in a biofilm, and/or in a community of different bacterial species is now possible using *gfp* constructs [22, 23].

The excitement associated with QS now is the possibility of designing new compounds or discovering natural compounds that will block QS activation of a key phenotypic trait, thus rendering a bacterium avirulent in a medical situation or unable to form a biofilm in either a medical or an industrial setting.

Acknowledgements

The knowledge presented above is primarily accumulated from the efforts of many researchers and students in the QS group at the University of Nottingham, UK. The efforts of Nigel Bainton, Mavis Daykin, Ram Chhabra, Leigh Fish, David Kirke, Martin Lynch, and Michael Winson and the guidance of Professors Barrie Bycroft, Paul Williams, and the late Gordon SAB Stewart are especially important to the author and are gratefully acknowledged.

Further reading

Dunny GM, Winans SC (eds) (1999) *Cell-cell signaling in bacteria*. ASM Press, Washington DC

Bainton NJ, Stead P, Chhabra SR et al. (1992) *N*-(3-oxohexanoyl)-L-homoserine lactone regulates carbapenem antibiotic production in *Erwinia carotovora*. *Biochem J* 288: 997–1004

Swift S, Winson MK, Chan PF et al. (1993) A novel strategy for the isolation of *luxI* homologues: evidence for the widespread distribution of a LuxR:LuxI superfamily in enteric bacteria *Mol Microbiol* 10: 511–520

Mayville P, Ji G, Beavis R et al. (1999) Structure-activity analysis of synthetic autoinducing thiolactone peptides from *Staphylococcus aureus* responsible for virulence. *Proc Natl Acad Sci USA* 1999 96: 1218–1223

Qazi SN, Counil E, Morrissey J et al. (2001) *agr* expression precedes escape of internalized *Staphylococcus aureus* from the host endosome. *Infect Immun* 69: 7074–7082

References

1   Schauder S, Bassler BL (2001) The languages of bacteria. *Genes Dev* 15: 1468–1480

2   Swift S, Downie JA, Whitehead NA et al. (2001). Quorum sensing as a population density dependent determinant of bacterial physiology. *Adv Microb Physiol* 45: 199–270

3   Fuqua WC, Winans SC, Greenberg EP (1994) Quorum sensing in bacteria- the LuxR-LuxI family of cell density-responsive transcriptional regulators. *J Bacteriol* 176: 269–275

4   Visick KL, Ruby EG (1999) The emergent properties of quorum sensing: consequences to bacteria of autoinducer signaling in their natural environment. In: GM Dunny, SC Winans (eds): *Cell-cell signaling in bacteria*. ASM Press, Washington DC, 333–352

5   McClean KH, Winson MK, Fish L et al. (1997) Quorum sensing and *Chromobacterium violaceum*: exploitation of viola-

cein production and inhibition for the detection of *N*-acylhomoserine lactones. *Microbiology* 143: 3703–3711

6   Winson MK, Swift S, Fish L et al. (1998) Construction and analysis of *luxCDABE*-based plasmid sensors for investigating *N*-acylhomoserine lactone-mediated quorum sensing. *FEMS Microbiol Letts* 163: 185–192

7   Swift S., Karlyshev AV, Fish L et al. (1997) Quorum sensing in *Aeromonas hydrophila* and *Aeromonas salmonicida*: identification of the LuxRI homologs AhyRI and AsaRI and their cognate *N*-acylhomoserine lactone signal molecules. *J Bacteriol* 179: 5271–5281

8   Laue BE, Jiang Y, Chhabra SR et al. (2000) The biocontrol strain *Pseudomonas fluorescens* F113 produces the *Rhizobium small* bacteriocin, *N*-(3-hydroxy-7-cis-tetradecenoyl)homoserine lactone, via HdtS, a putative novel *N*-acylhomoserine lactone synthase. *Microbiology* 146: 2469–2480

9   Pearson JP, Gray KM, Passador L et al. (1994) Structure of the autoinducer required for expression of *Pseudomonas aeruginosa* virulence genes. *Proc Natl Acad Sci USA* 91: 197–201

10 Shaw PD, Ping G, Daly SL et al. (1997) Detecting and characterizing *N*-acyl-homoserine lactone signal molecules by thin-layer chromatography. *Proc Natl Acad Sci USA* 94: 6036–6041

11 Camara M, Daykin M, Chhabra SR (1998) Detection, purification and synthesis of *N*-acyl homoserine lactone quorum sensing molecules. *Methods Microbiol* 27: 319–330

12 Holden MT, Chhabra SR, de Nys R et al. (1999) Quorum-sensing cross talk: isolation and chemical characterization of cyclic dipeptides from *Pseudomonas aeruginosa* and other gram-negative bacteria. *Mol Microbiol* 33: 1254–1266

13 Charlton TS, de Nys R, Netting A et al. (2000) A novel and sensitive method for the quantification of *N*-3-oxoacyl homoserine lactones using gas chromatography-mass spectrometry: application to a model bacterial biofilm. *Environ Microbiol* 2: 530–541

14 Schaefer AL, Greenberg EP, Parsek MR (2001) Acylated homoserine lactone detection in *Pseudomonas aeruginosa* biofilms by radiolabel assay. *Methods Enzymol* 336: 41–47

15 Rodelas B, Lithgow JK, Wisniewski-Dye F et al. (1999) Analysis of quorum-sensing-dependent control of rhizosphere-expressed (*rhi*) genes in *Rhizobium leguminosarum* bv. *viciae*. *J Bacteriol* 181: 3816–3823

16 Pearson JP, Pesci EC, Iglewski BH (1997) Roles of *Pseudomonas aeruginosa las* and *rhl* quorum-sensing systems in control of elastase and rhamnolipid biosynthesis genes. *J Bacteriol* 179: 5756–5767

17 Milton DL, Chalker VJ, Kirke D et al. (2001) The LuxM homologue VanM from *Vibrio anguillarum* directs the synthesis of *N*-(3-hydroxyhexanoyl)homoserine lactone and *N*-hexanoylhomoserine lactone. *J Bacteriol* 183: 3537–3547

18 Swift S, Lynch MJ, Fish L (1999) Quorum sensing dependent regulation and blockade of oxoprotease production in *Aeromonas hydrophila*. *Infect Immun* 67: 5192–5199

19 Lindum PW, Anthoni U, Christophersen C et al. (1998) *N*-Acyl-L-homoserine lactone autoinducers control production of an extracellular lipopeptide biosurfactant required for swarming motility of *Serratia liquefaciens* MG1. *J Bacteriol* 180: 6384–6388

20 Bassler BL, Greenberg EP, Stevens AM. (1997) Cross-species induction of luminescence in the quorum-sensing bacterium *Vibrio harveyi*. *J Bacteriol* 179: 4043–4045

21 Baca-DeLancey RR, South MM, Ding X, Rather PN (1999) *Escherichia coli* genes regulated by cell-to-cell signaling. *Proc Natl Acad Sci USA* 96: 4610–4614

22 Andersen JB, Heydorn A, Hentzer M et al. (2001) *gfp*-based *N*-acyl homoserine-lactone sensor systems for detection of bacterial communication. *Appl Environ Microbiol* 67: 575–585

23 Qazi SN, Counil E, Morrissey J et al. (2001) *agr* expression precedes escape

of internalized *Staphylococcus aureus* from the host endosome. *Infect Immun* 69: 7074–7082

24 Thomson NR, Crow MA, McGowan SJ et al. (2000) Biosynthesis of carbapenem antibiotic and prodigiosin pigment in *Serratia* is under quorum sensing control. *Mol Microbiol* 36: 539–556

25 Pierson III LS, Pierson EA (1996) Phenazine antibiotic production in *Pseudomonas aureofaciens*: role in rhizosphere ecology and pathogen suppression. *FEMS Microbiol Lett* 136: 101–108

26 Stewart GSAB, Lubinsky-Mink S, Jackson CG et al. (1986) pHG165: a pBR322 copy number derivative of pUC8 for cloning and expression. *Plasmid* 15: 172–181

27 Keen NT, Tamaki S, Kobayashi D, Trollinger D (1988) Improved broad-host-range plasmids for DNA cloning in gram-negative bacteria. *Gene* 70: 191–197

28 Bartolome B, Jubete Y, Martinez E, de la Cruz F (1991) Construction and properties of a family of pACYC184-derived cloning vectors compatible with pBR322 and its derivatives. *Gene* 102: 75–78

11 Transcriptional Profiling in Bacteria Using Microarrays

Michael T. Laub and R. Frank Rosenzweig

Contents

1 Introduction

Recent advances in robotics and computer science now make possible the acquisition and processing of immense amounts of genetic data. Because of their small size and relative simplicity, microbial genomes were the proving grounds for technologies that accelerated the Human Genome Project [1–4]. The present state of these technologies is such that it is feasible to completely sequence a microbial genome in days rather than years. As of the time this volume went to press, more than 60 microbial genomes had been sequenced and provisionally annotated (http://www.tigr.org/tigr-scripts/CMR2/CMRHome-Page.spl).

Methods and Tools in Biosciences and Medicine
Prokaryotic Genomics, ed. by M. Blot
© 2003 Birkhäuser Verlag Basel/Switzerland

Complete genome sequence data have spawned novel, holistic approaches to studying biological systems that will transform the use of traditional reductionist approaches in the next century. Already, such terms as "transcriptome," "proteome", and "metabolome" have become well established within the scientific vernacular. The transcriptome became experimentally accessible through the pioneering efforts of Fodor, Davis, Brown and colleagues. These efforts focused on synthesizing *in situ* [5] or depositing and covalently linking [6] thousands of known deoxyribonucleotides in massively parallel fashion onto a solid surface. The resulting DNA microarray serves as a platform for global analyses of the transcriptome (expression profiling) and genome (comparative genomic hybridization).

Using a yeast model, DeRisi et al. [7] demonstrated the power of the microarray approach for investigating how the transcriptome is reprogrammed in response to externally or internally imposed variables. Yeast isolated at various stages of growth in glucose batch culture (e. g., exponential phase, diauxic shift) were harvested and processed for mRNA. Each of these messenger RNA pools was then reverse-transcribed in the presence of dNTPs, one of which had been labeled with a fluorescent dye (Cy5), resulting in Cy5-labeled cDNAs. Reference cDNAs were derived from mRNAs isolated in the earliest stage of growth; these were reverse-transcribed in the presence of dNTPs labeled with a different fluor (Cy3). An equimolar mixture of each stage-specific Cy5-labeled cDNA pool was co-hybridized on a microarray with the Cy3-labeled reference (Fig. 1). This procedure gave rise to a temporal profile of changes in

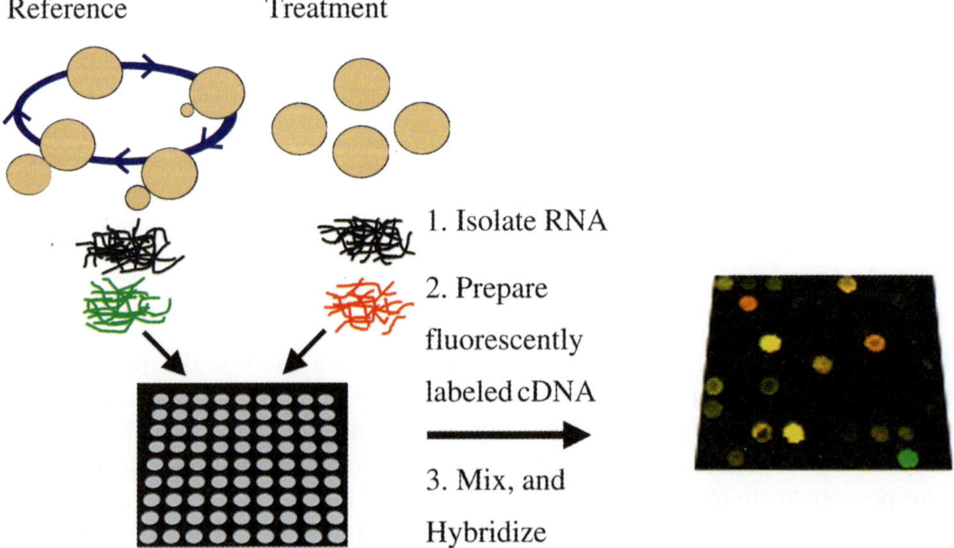

Figure 1   The microarray procedure.
The experimental objective in this example is to compare the transcriptional profile of cells in one growth phase (Treatment) to that of mixed-phase cells (Reference). (Figure courtesy of D. Botstein, Stanford University).

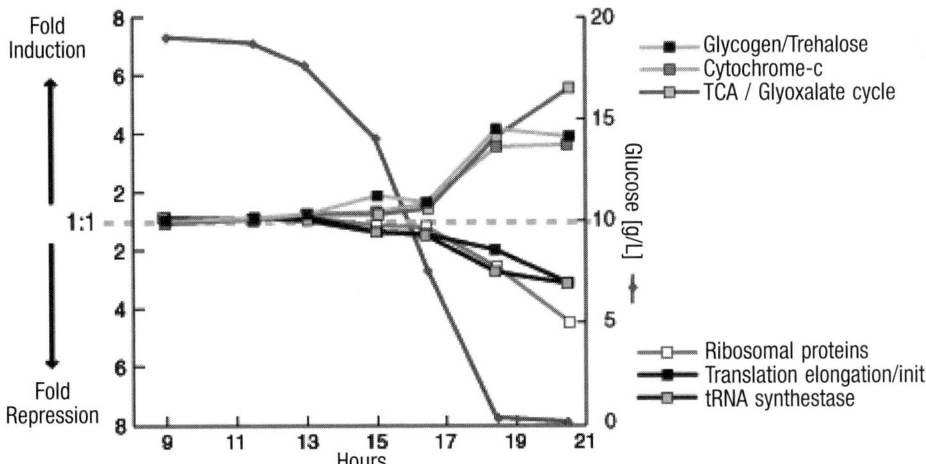

Figure 2 Transcriptional profiling in yeast during the course of glucose batch culture reveals coordinate expression of different sets of genes [7].

the expression of each gene in the yeast genome during extended batch culture (Fig. 2). In the same study, DeRisi et al. investigated the impact of mutations in key regulatory genes known to influence gene expression in glucose-sufficient batch cultures. This complementary approach – investigating system response first to external and then to internal state variables – has become an experimental paradigm in DNA microarray studies [8, 9].

The microbiological literature is being increasingly enriched by a growing number of studies that use microarray analysis. The technique has been applied successfully to diverse species that include gram-negative Eubacteria *Escherichia coli* [10–12] and *Pseudomonas aeruginosa* [13], gram-positive Eubacteria *Bacillus subtilis* [14] and *Staphylococcus aureus* [15] also see [16], as well as the Archaeon *Pyrococcus furiosus* [17]. Indeed, any microbial taxon for which there exist reasonably complete genome sequence data is a candidate for this mode of analysis. Global gene-expression profiles have been established for free-living phototrophs [18], heterotrophs [19], and both obligate [20] and opportunistic pathogens [13]. Collectively, these studies have yielded fundamentally new insight into regulatory circuits that govern processes such as quorum sensing [21, 22], progress through the cell cycle [19] and global responses to stress in the forms of heat [23, 24], peroxide [8], and nitrogen limitation [9].

DNA microarray analysis invariably enlarges the set of genes thought to be coordinately regulated in response to the experimental variable of interest. Peterson et al. (2000), investigating competence stimulating pheromone (CSP) regulation in *Streptococcus pneumoniae*, found eight additional loci not previously associated with competence that could be induced by CSP [21]. Characteristically, the set is enlarged with genes having no known function (*FUN* genes). Investigating genomic responses to low iron in *Pasteurella multocida*, Paustian et al. (2001) determined that 27% of genes with significantly altered expression fell into this category [25].

In this regard, DNA microarray analysis provides a powerful tool for discovering gene function [16]. By linking databases that describe the transcriptome's response to various environmental and genetic perturbations, it will be possible to understand how such genes fit into cellular regulatory circuits. Such understanding will drive formulation of hypotheses that can be tested by reductionist methods. This procedure will permit strong inference concerning gene function. The Gene Ontology project at Stanford University (http://www.geneontology.org) is a model database designed for assimilating such information in Eukaryotic systems. The Comprehensive Microbial Resource homepage maintained by The Institute for Genomic Research (TIGR) is a model resource that serves the Prokaryotic community http://www.tigr.org/tigr-scripts/CMR2/CMRHomePage.spl.

The work of Schut et al. [17] is particularly instructive in this regard. Using a subset of ORFs from the *Pyrococcus furiosus* genome, these workers investigated the effects of elemental sulfur (S°) on gene expression at 95 °C using maltose as sole carbon source. Among the uncharacterized ORFs were two genes whose expression profiles and sequence homologies suggest they are part of a novel S°-reducing, membrane-associated iron-sulfur cluster-containing complex. This finding can be further investigated by using conventional biochemical and genetic techniques. Clearly, progress in microbial genome annotation will be an iterative process driven by homology search of expanding databases that link microarray data with information obtained by traditional methods.

Although it is beyond the scope of this chapter, we should note that DNA microarrays provide a platform for doing comparative genomics among related taxa or among isolates belonging to the same "species" [26–28]. These analyses are based on co-hybridization of differently labeled genomic DNA isolated from test and reference taxa. Prior to labeling, these gDNAs are sheared or restriction-digested, resulting in a homogenous pool of sub-kilobase fragments. Hybridization of these fragments to DNA microarrays enables the investigator to analyze architectural features of the genome such as linkage, map position, and copy number. Comparative genomic hybridization has revealed novel linkage groups related to virulence in *Shigella* [29], astonishing genome plasticity in *Helicobacter pylori* [30], and compelling evidence for widespread lateral gene transfer among *Salmonella* and *Streptococcus* serovars [31, 32].

The present chapter aims to provide a set of genomic protocols and Web-based resources for conducting global expression studies in microbial systems. The authors' expertise lies in the use of spotted DNA microarrays [19, 33, 34] that consist of full or partial ORFs. Consequently, our protocols and our discussion will be restricted to these tools. Information on alterative technologies is presented in section 5, Web-based resources.

2   Materials

The following protocol describes the execution of an expression profiling experiment in which DNA microarrays are used to probe relative mRNA levels within a bacterial cell. However, as microarrays are premised on basic nucleic acid hybridization principles, this protocol can be modified to accommodate any input DNA or RNA sample. For instance, DNA microarrays have been used to monitor the status of DNA replication [35], to compare the DNA content of various bacterial strains [36], and to detect DNA-protein interactions [34].

2.1   Strains

Expression profiling has been successful for a wide range of bacterial species, including species that are genetically or experimentally intractable. Microarrays are thus a general tool with the only requirement being the ability to harvest sufficient quantities of RNA. Investigators should be aware that in 2002 the National Institute of Allergy and Infectious Diseases (NIAID) established a Pathogen Functional Genomics Resource Center through TIGR (http://www.pfgrc.tigr.org). Initially, TIGR will develop genomic resources for three organisms, *Streptococcus pneumoniae*, *Staphylococcus aureus*, and *Salmonella typhimurium*. These resources will consist of the type strain, genomic DNA, PCR primer pair sets, DNA glass slide microarrays, and full-length entry clone sets. By 2006, the NIAID initiative will result in genomic resources for 10 microbial pathogens.

2.2   Microarrays

There are two general types of microarrays available for expression profiling work. Oligonucleotide arrays (such as those commercially available from Affymetrix) consist of short, typically 20–25 nucleotide, single-stranded oligomers arrayed on a solid substrate. These arrays are usually manufactured by synthesis of the oligomers directly on the array substrate. The second type are generally referred to as spotted cDNA arrays. The spots on these arrays are usually double-stranded PCR products, the result of amplifying specific regions of chromosomal DNA or clones of a cDNA library. This latter type of array can be relatively easily and inexpensively fabricated in a typical academic laboratory or purchased from a variety of commercial sources.

2.3   Scanner

The fluorescent read-out of a hybridized microarray requires a scanner. There are a wide range of laser-based scanners available, but in head-to-head comparisons, they have been shown to give nearly identical results [37].

3   Methods

Protocol 1   RNA preparation

1. Pellet cells, pour off supernatant, then resuspend in 500 μl TE+0.5 mg/ml lysozyme.
2. Add 50 μl 10% SDS, vortex for 10 s, place in 64 °C water bath for 2 min.
3. Add 55 μl 1M NaAcetate, pH 5.2.
4. Add 500 μl H_2O-saturated acid phenol (pH < 7.0), place at 64 °C for 5 min then shake/vortex periodically.
5. Centrifuge for 10 min. at > 10,000 rpm.
6. Remove aqueous layer into new tube and add equal volume chloroform.
7. Centrifuge for 10 min at > 10,000 rpm.
8. Remove aqueous layer into new tube, add 1/10 volume 3M NaAcetate, 1 mM EDTA.
9. Add 2.0–2.5 volumes 100% cold ethanol, place at –20 °C for > 20 min.
10. Centrifuge for 25 min at > 10,000 rpm.
11. Wash pellet once with 70% cold ethanol.
12. Pour off ethanol, air dry pellet and resuspend in H_2O.
13. Degrade any residual genomic DNA by treating pellet with RNase-free DNase (add 2 U DNase per 100 μg RNA, incubate at 37 °C for 1 h, and stop reaction by adding 0.5M EDTA).
14. Reisolate RNA by phenol/chloroform extraction (use acid phenol as above).
15. Precipitate RNA as above and resuspend in H_2O.

Protocol 2   cDNA synthesis and labeling

1. Mix 13 µl RNA (25–35 µg total RNA) with 1 µl (5 µg) random hexamers/octamers + 1 µl (2 µg) gene-specific primers.
2. Incubate at 65 °C for 10 min to denature RNA, then place on ice.
3. Add 3 µl Cy5-dCTP to one sample, 3 µl Cy3-dCTP to other sample.
4. Add 3 µl DTT, 6 µl 5X 1ˢᵗ strand buffer.
5. add 0.6 µl dNTP mix (25 mM dATP, 25 mM dGTP, 25 mM dTTP, 10 mM dCTP).
6. Add 2 µl reverse transcriptase (Superscript II – GibcoBRL).
7. Place at room temperature for 10 min.
8. Incubate at 42°C for 110 min.
9. Stop reaction and degrade RNA by adding 1.5 µL 1N NaOH and placing at 65 °C for 10 min.
10. Add 1.5 µl 1N HCl to neutralize.
11. Purify probe (3X spins through a size-fractionation column such as Micro-con-30 columns or PCR purification kits such as Qiagen). The final volume of purified probe should be 14 µl.
12. Add 1.5 µl yeast tRNA (5 µg/µl) as blocking reagent, 2 µl 20X SSC, and 0.25 µl 10% SDS.
13. Boil 1 min, let cool to room temperature, apply directly to array, and cover with 22 × 22 mm coverslip (adjust all volumes proportionally for larger arrays/coverslips).
14. Place array inside airtight chamber and place chamber in 65 °C water bath for 8–12 h.

Protocol 3   Hybridization and scanning

1. Three wash solutions are used: I=1X SSC/0.03% SDS, II=0.2X SSC, and III=0.05X SSC.
2. Disassemble chamber, place array in slide rack immersed in wash solution I for 1 min, and agitate rack in solution, minimizing air exposure.
3. Transfer to wash solution II for 1 min with agitation.
4. Transfer to wash solution III for 1 min with agitation.
5. Spin slide rack for 5 min at 500 rpm.
6. Scan array using fluorescent microarray scanner (see notes above).
   Image analysis and data processing
   Mask array: a variety of software tools exist for identifying spots on a fluorescent array image and then calculating values for each spot such as signal and background. These software tools also typically facilitate the process of mapping fluorescent spots to gene names. ScanAlyze is a popular, freely distributed software tool for these tasks (see http://rana.lbl.gov).

4   Troubleshooting

Although the protocol specifies using 25–35 μg of RNA, smaller amounts can be used if starting material is limiting. There are several modifications that can be made to accommodate for less starting material. One popular modification is to increase the ratio of fluorescent nucleotide to non-fluorescent nucleotide, thereby biasing the reaction to incorporate more dye. There are also methods for the amplification of small quantities of RNA [38, 39].

As bacterial mRNAs generally do not have polyA tails, e. g., eukaryotic mRNAs, the starting material for an expression profiling experiment is simply total RNA. This has led to concerns that the ribosomal RNA, constituting as much as 98% of a typical pool of total RNA, will interfere somehow with hybridization or give rise to a high background. While protocols have been developed to subtract rRNA from total RNA, such measures do not appear necessary. Total RNA has been successfully used for expression analysis, suggesting that labeled rRNA is sufficiently outcompeted during hybridization and then removed during the wash steps.

For priming the RT reaction, either random primers or gene-specific primers can be used, or a combination of both. For completely sequenced organisms, an algorithm has been developed that allows one to design the minimal set of primers that will anneal to and prime the RT reaction from all genes [40].

It has been well documented that nucleotides conjugated to bulky fluorescent dyes can interfere with incorporation by the reverse transcriptase. Further, it has been shown that the Cy3- and Cy5-labeled nucleotides incorporate with different efficiencies. These problems can be circumvented by incorporating an amino-allyl nucleotide during the reverse transcriptase reaction and then subsequently coupling reactive Cy3 and Cy5 dyes to the allyl group. Because the same amino-allyl nucleotide is incorporated into the two samples to be compared, the incorporation efficiency is identical. While this eliminates the problem of differential incorporation efficiency, it should be noted that the only amino-allyl nucleotide available is dUTP. For highly GC-rich genomes, the signal intensity obtained by indirect incorporation of amino-allyl dUTP is generally insufficient, making direct incorporation of fluorescently tagged dCTP the better option, despite the differences in incorporation efficiency.

Undoubtedly, the most complicated and time-consuming aspect of expression analysis is the data analysis [41]. After scanning an array and generating the fluorescent images, the investigator must process the array images to remove inconsistent or abnormal spots, filter out spots whose intensity is too close to background to allow reliable determination of a ratio, and normalize the intensity of each fluorescent channel. Individual arrays also must be combined with other repetitions of the same experiment/comparison to allow a statistical analysis and an assessment of the significance of any gene expression changes observed (Fig 3 and 4). Finally, the comparison of different experiments, such as a comparison of different mutant strains,

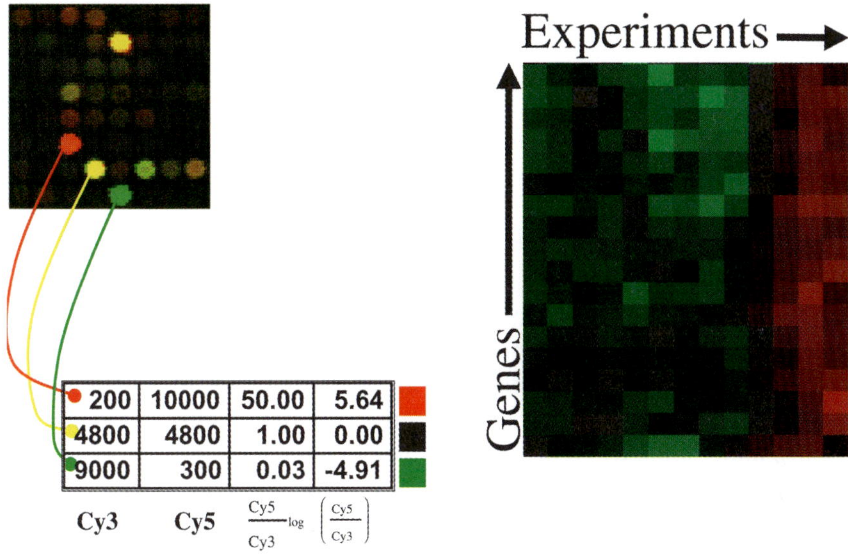

$$\frac{Cy5}{Cy3} \qquad \log\left(\frac{Cy5}{Cy3}\right)$$

Figure 3 Extracting data. Slides are scanned at the appropriate excitation/ emisson spectra and intensities recorded in dye-specific channels. Log$_2$-ratio intensities reveal fold-differences between Reference (green) and Treatment (red). These are color-coded and presented in a GENES * EXPERIMENT matrix (Figure courtesy of D. Botstein, Stanford Univ.).

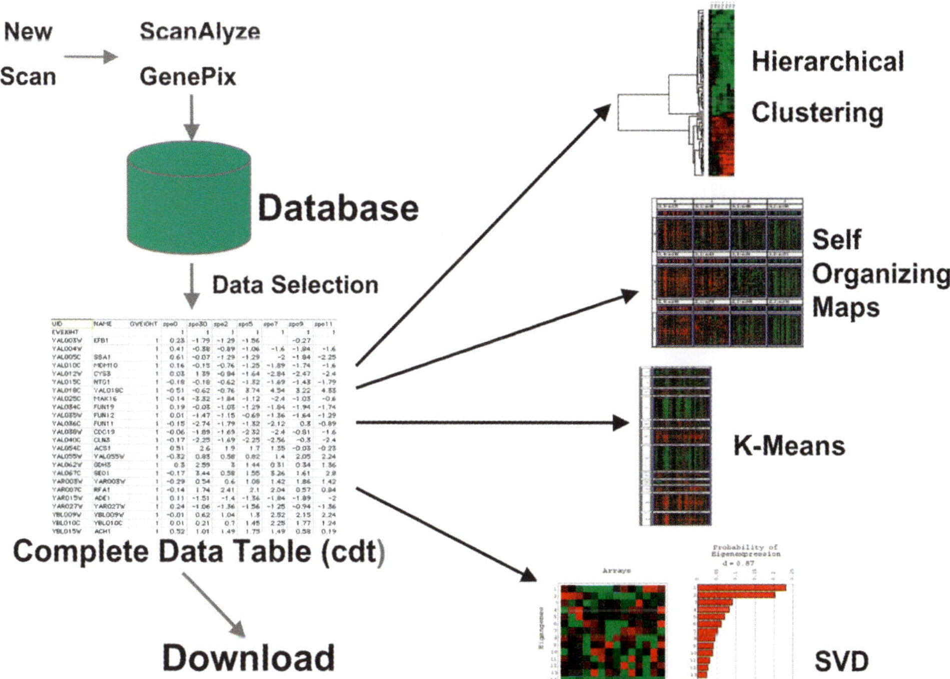

Figure 4 Data flow in microarray studies. Following laser scanning, data are entered into a Complete Database Table (cdt). Multiple software packages are available as freeware for post-scan analyses (Figure courtesy of D. Botstein, Stanford University).

involves an enormous amount of data management and data mining activities. A full description of all of these issues is beyond the scope of this article, but these are reviewed in depth in [42].

5   Web-based microarray resources

The World Wide Web offers numerous resources to assist investigators in the design, execution, and analysis of DNA microarray experiments. These resources also offer advice on acquiring – and even constructing – the necessary equipment for conducting such experiments. We have compiled a brief list of sites whose many links will enable the reader to build a more comprehensive list of resources. We applaud the authors of these sites – and many others not listed – for their commitment to freely disseminate new ideas and techniques. That commitment is consistent with the spirit captured by Claude Bernard's proverb, "Art is I, and science is we."

5.1   Databases

- http://www.tigr.org/tigr-scripts/CMR2/CMRHomePage.spl
- http://www.geneontology.org
- http://www.jgi.doe.gov/JGI_microbial/html/index.html

5.2   Experimental design

- http://www.microarrays.org
  DeRisi Lab-UCSF
- http://cmgm.stanford.edu/pbrown/mguide/index.html
  Brown Lab-Stanford

5.3   Microarray forum

- http://cmgm.stanford.edu/cgi-bin/cgiwrap/taebshin/dcforum/dcboard.cgi

5.4   Custom arraying

- http://www.genetics.ucla.edu/microarray/CustomArray.htm

5.5   Commercial arrays

* http://www.affymetrix.com/products/index.affx

5.6   Software and data analysis

* http://genome-www5.stanford.edu/MicroArray/SMD/restech.html
* http://ihome.cuhk.edu.hk/~b400559/arraysoft.html
* http://rana.lbl.gov

6   Acknowledgements

The authors wish to acknowledge support from the Department of Energy's Microbial Cell Project (to ML) and the National Institutes of Health [GM63800–01] and the National Science Foundation's EPSCoR Program [EPS-0091995] (to FR). Figures 1, 3 and 4 were generously provided by Professor David Botstein of Stanford University School of Medicine. We wish to thank Margie Kinnersley for helpful discussion and assistance in preparing this manuscript. Protocols 3.2 and 3.3 are derived from those originally developed by Arkady Khodursky of the University of Minnesota.

References

1 Bult CJ, Whit O, Olsen GJ, Zhou L, et al (1996) Complete genome sequence of the methanogenic archaeon, *Methanococcus jannaschii. Science* 273: 1058–1073

2 Fraser CM, Gocayne JD, White O, Adams MD, et al (1995) The minimal gene complement of *Mycoplasma genitalium. Science* 270: 397–403

3 Goffeau, A et al. (1997) The yeast genome directory. *Nature* 387: suppl. 5–6.

4 International Human Genome Mapping Consortium. 2001. A physical map of the human genome. *Nature* 409: 934–941

5 Fodor SP, Rava RP, Huang XC, Pease AC, et al (1993) Multiplexed biochemical assays with biological chips. *Nature* 364: 555–556

6 Schena M, Shalon D, Davis RW, Brown PO (1995) Quantitative monitoring of gene expression patterns with a complementary DNA microarray. *Science* 270: 467–470

7 DeRisi JL, Iyer VR, Brown PO (1997) Exploring the metabolic and genetic control of gene expression on a genomic scale. *Science* 278: 680–686

8 Ming Zheng XW, Templeton LJ, Smulski DR, LaRossa RA et al (2001) DNA microarray-mediated transcriptional profiling of the *Escherichia coli* response to hydrogen peroxide. *J Bacteriol* 183: 4562–4570

9 Zimmer DP, Soupene E, Lee HL, Wendisch VF (2000) Nitrogen regulatory protein C-controlled genes of *Escherichia coli*: Scavenging as a defense against nitrogen limitation. *Proc Nat Acad Sci* (USA) 97 (26): 14674–14679

10 Blattner FR, Richmond CS, Glasner JD, Mau R et al (1999) Genome-wide expression profiling in *Escherichia coli* K-12. *Nuc Acids Res* 27 (19): 3821–3835

11 Tao Han CB, Richmond C, Blattner FR, Conway T (1999) Functional genomics: Expression Analysis of *Escherichia coli* growing on minimal and rich media. *J Bacteriol* 181: 6425–6440

12 Wei Y, Lee J, Smulski DR, LaRossa RA (2001) Global impact of sdiA amplification revealed by comprehensive gene expression profiling of *Escherichia coli. J Bacteriol* 183: 2265–2272

13 Greenberg EP, Whiteley M, Bangera MG, Bumgarner RE et al (2001) Gene expression in *Pseudomonas aeruginosa* biofilms. *Nature* 413 (6858): 860–864

14 Tanaka T, Ogura M, Yamaguchi H, Yoshida K et al (2001) DNA microarray analysis of *Bacillus subtilis* DegU, ComA and PhoP regulons: An approach to comprehensive analysis of *B. subtilis* two-component regulatory systems. *Nuc Acids Res* 29 (18): 3804–3813

Hoch JA, and M Perego (2001) Functional Genomics of Gram-Positive Microorganisms: Review of the Meeting, San Diego, California, 24 to 28 June 2001. *J Bacteriol* 183: 6973–6978

Kobayashi K, Ogura M, Yamaguchi H, Yoshida K, Ogasawara N, Tanaka T, Fujita Y (2001) Comprehensive DNA microarray analysis of *Bacillus subtilis* two-component regulatory systems. *J Bacteriol* 183: 7365–7370

Ye RW, Tao W, L Bedzyk L, Young T, Chen M, Li L (2000) Global gene expression profiles of *Bacillus subtilis* grown under anaerobic conditions. *J Bacteriol* 182: 4458–4465

15 Dunman PM, Murphy E, Haney S, Palacios D et al (2001) Transcription profiling-based identification of *Staphylococcus aureus* genes regulated by the agr and/or sarA loci. *J Bacteriol* 183(24): 7341–7353

16 Hughes TR, Marton MJ, Jones AR, Roberts CJ et al (2000) Functional discovery *via* a compendium of expression profiles. *Cell* 102: 109–126

17 Schut GJ, Zhou J, Adams MW (2001) DNA microarray analysis of the hyperthermophilic archaeon *Pyrococcus furiosus*: Evidence for a New type of sulfur-reducing enzyme complex. *J Bacteriol* 183: 7027–7036

18 Singh AK, Sherman LA (2000) Identification of iron-responsive, differential gene expression in the cyanobacterium *Synechocystis* sp. Strain PCC 6803 with a customized amplification library. *J Bacteriol* 182: 3536–3543

19 Laub MT, McAdams HH, Feldblyum T, Fraser CM et al (2000) Global analysis of the genetic network controlling a bacterial cell cycle. *Science* 290: 2144–8

20 Ang S, Lee ZC, Peck K, Sindici M et al (2001) Acid-induced gene expression in *Helicobacter pylori*: study in genomic scale by microarray. Infect Immun 69:1679–76.

21 Peterson S, Cline RT, Tettelin H, Sharov V, et al (2000) Gene expression analysis of the *Streptococcus pneumoniae* competence regulons by use of DNA microarrays. *J Bacteriol* 182: 6192–6202

22 de Saizieu A, Gardes C, Flint N, Wagner C et al (2000) Microarray-based identification of a novel *Streptococcus pneumoniae* regulon controlled by an autoinduced peptide. *J Bacteriol* 182: 4696–4703

23 Helmann JD, Wu MF, Kobel PA, Gamo FJ et al (2001) Global transcriptional response of *Bacillus subtilis* to heat shock. *J Bacteriol* 183: 7318–7328

24 Smoot LM, Smoot JC, Graham MR, Somerville GA et al (2001) Global differential gene expression in response to growth temperature alteration in group A Streptococcus. *Proc Natl Acad Sci (USA)* 98: 10416–21

25 Paustian ML, May BJ, Kapur V (2001) *Pasteurella multocida* gene expression in response to iron limitation. *Infect Immun* 69: 4109–15

26 Fitzgerald JR, Musser JM (2001) Evolutionary genomics of pathogenic bacteria. *Trends Microbiol* 9: 547–553

27 Kato-Maeda M, Rhee JT, Gingeras TR, Salamon H, et al (2001) Comparing genomes within the species *Mycobacterium tuberculosis*. *Genome Res* 11: 547–54

28 Malloff CA, Fernandez RC, Lam WL (2001) Bacterial comparative genomic hybridization: a method for directly identifying lateral gene transfer. *J Mol Biol* 312: 1–5

29 Day WA Jr, Maurelli AT (2001) Pathoadaptive mutations that enhance virulence: Genetic organization of the cadA regions of *Shigella* spp. *Infect Immun* 69: 7471–7480

30 Salama N, Guillemin K, McDaniel TK, Sherlock G et al (2000) A whole-genome microarray reveals genetic diversity among *Helicobacter pylori* strains. *Proc Natl Acad Sci (USA)* 97: 14668–14773

31 Hakenbeck R, Balmelle N, Weber B, Gardes C et al (2001) Mosaic genes and mosaic chromosomes: intra- and interspecies genomic variation of *Streptococcus pneumoniae*. *Infec Immun* 69: 2477–2486

32 McClelland M, Sanderson KE, Spieth J, Clifton SW et al (2001) Complete genome sequence of *Salmonella enterica* serovar Typhimurium LT2. *Nature* 413: 852–856

33 Ferea T, Botstein D, Brown PO, Rosenzweig RF (1999) Systematic changes in gene expression patterns following adaptive evolution in yeast. *Proc Natl Acad Sci (USA)* 96: 9721–9726

34 Laub MT, Chen SL, Shapiro L, McAdams HH (2002) Genes directly controlled by CtrA, a master regulator of the *Caulobacter* cell cyclo. *Proc Natl Acad Sci (USA)* 99: 4632–4637

35 Khodursky AB, Peter BJ, Schmid MD, DeRisi J et al (2000) Analysis of topoisomerase function in bacterial replication fork movement: use of DNA microarrays. *Proc Natl Acad Sci (USA)* 97: 9419–9424

36 Behr MA, Wilson MA, Gill WP, Salamon H et al (1999) Comparative genomics of BCG vaccines by whole-genome DNA microarray. *Science* 284: 1520–1523

37 Ramdas L, Wang J, Hu L, Cogdell D (2001) Comparative evaluation of laser-based microarray scanners. *Biotechniques* 31: 546, 548, 550, *passim*

38 Eberwine J, Yeh H, Miyashiro K, Cao Y et al (1992) Analysis of gene statement in single live neurons. *Proc Natl Acad Sci (USA)* 89: 3010–3014

39 Phillips J, Eberwine JH (1996) Antisense RNA amplification: A linear amplification method for analyzing the mRNA

population from single living cells. *Methods Enzymology* 10: 283–288

40 Talaat AM, Hunter P, Johnston SA (2000) Genome-directed primers for selective labeling of bacterial transcripts for DNA microarray analysis. *Nat Biotechnol* 18: 679–682

41 Sherlock G (2000) Analysis of large-scale gene expression data. *Curr Opinion Immun* 12: 201–205

42 Quackenbush J (2001) Computational genetics and computational analysis of microarray data. *Nat Rev Genet* 2: 418–427

12 Transcriptome Analysis by Macroarrays

Cécile Jourlin-Castelli, François Denizot and Philippe Bouloc

Contents

1 Introduction

A new challenge of prokaryotic biology is to acquire a comprehensive view of regulatory networks and to appreciate their interconnections. Transcriptome technologies are used to generate a quantified profile of transcriptional expression and are applicable to any organism for which DNA sequence information is available [1]. The transcriptional consequences of a given growth condition or mutation can be exhaustively compared with reference situations [2]. For example, cells growing in nutrient-poor medium showed increased expression of amino acid biosynthetic genes that are repressed in a

Methods and Tools in Biosciences and Medicine
Prokaryotic Genomics, ed. by M. Blot
© 2003 Birkhäuser Verlag Basel/Switzerland

rich medium [3]. Adaptations to stress situations, such as changes in temperature or osmotic pressure or the presence of toxic compounds, require an extensive reorganization of gene expression, resulting in a specific transcriptome profile. These genome-wide profiles can be used to dissect the transcriptional regulatory circuitry and to determine which genes are coordinately regulated. Enhanced or reduced transcription of genes in response to the modification of a specific regulatory protein will allow the identification of all genes belonging to a given regulon. Genes responding to a specific external signal (stimulus), such as a modification of environmental conditions, will lead to the identification of genes belonging to a stimulon. Transcriptome technology is straightforward and can reduce the need for long genetic analyses for the identification of regulatory pathways. However, transcriptional analysis does not take into account translational and post-translational regulation. Proteome studies might provide in theory a more complete view of cell adaptation to given conditions, but in practice only a limited number of proteins can be detected. In addition, the determination of a transcriptome profile is at present an easier and more discriminating technique when matrix arrays are available.

The general principle of transcriptome technology is based on the hybridization of cDNA (prepared from RNA extracted from the studied cells) to a matrix containing ordered spots of DNA fragments corresponding to the genes of interest (see Fig. 1). The ratio of hybridization intensity of each spot in test *versus* control samples is determined. Two main techniques are commonly used: 1) macroarray format on membranes, for use with radiolabeled probes, which will be presented in this chapter; or 2) microarray format on glass slides, for use with fluorescently labeled probes, which is presented in Chapter 11.

Macroarrays are usually prepared on nylon or polypropylene filters, on which around 10 ng per DNA fragment is spotted at defined positions. Membranes containing spotted DNA fragments corresponding to all putative genes are commercially available for prokaryotic genomes such as *Escherichia coli* or *Bacillus subtilis*. The spotted DNA fragments can be either PCR products, often corresponding to a complete or a portion of an ORF, or long oligonucleotides typically ranging from 40 to 70 mers.

The first step of a transcriptome analysis involves preparation of total RNA from strains cultivated under both test and control conditions (or using a mutated strain and the isogenic control strain). mRNA, which is very unstable, is converted to cDNA (copy-DNA) by a reverse transcriptase, making a stable copy of mRNAs; during this step, newly synthesized cDNA is radioactively labeled. RNA is removed afterwards. The cDNA is therefore a copy that represents the mRNA pool of a given condition. Upon incubation of the prepared cDNA with the macroarray, the cDNA molecules will bind to their complementary counterparts fixed on the macroarray. Hybridization is detected by radioactive signals. The radioactively labeled macroarray is exposed to storage phosphorimager screens having a high sensitivity and a wide linear dynamic range for signal detection. Radioactivity per spot is read

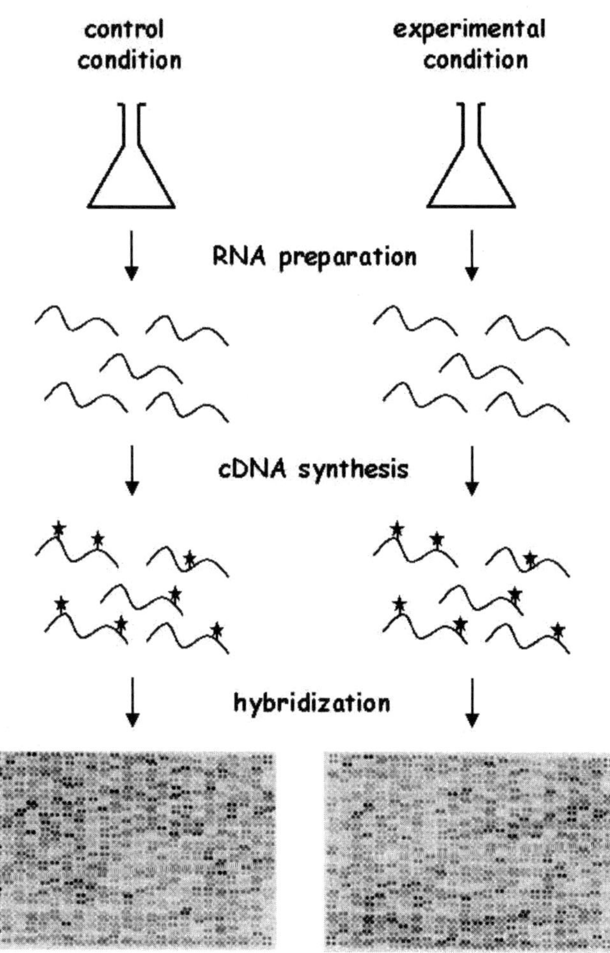

Figure 1 General scheme of transcriptome experiments. Images come from experiments performed using *E. coli* macroarrays manufactured by Sigma Genosys.

out from the screen on a phosphorimager screen reader, and the resulting image is processed by dedicated array analysis softwares. The latter will generate numeric tables where each gene, represented by a spotted DNA fragment, corresponds to a value proportional to a radioactive signal. Differences in signal intensities among the represented genes in the test state compared to the control state will reveal the set of genes that are subject to transcriptional modification.

2   Materials

2.1 Equipment

For macroarray hybridization the following equipment is recommended:
- Roller bottles
- Hybridization oven

2.2 Reagents

- DiMethyl Di-Carbonate (DMDC) supplied by Fluka.
- Absolve® is supplied by Dupont de Nemours.

2.3 Buffers

- Solution 1: sodium acetate 0.02 M, SDS 0.5%, EDTA 1 mM, pH 5.5.
- 5X SSPE (solution 2): 0.9 M NaCl, 50 mM sodium phosphate pH 7.7, and 5 mM EDTA.
- Hybridization solution (solution 3): 5X SSPE, 2% sodium dodecyl sulfate (SDS), 1X Denhardt's reagent, 100 μg/ml of sheared salmon sperm DNA.
- Washing solution (solution 4): 0.5X SSPE, 0.2% SDS.
- Stripping solution (solution 5): 10 mM Tris pH 7.5/8.0, 1 mM EDTA, 1% SDS.

3   Methods

Protocol 1   Treatment of material used for RNA preparation

1. To avoid any RNase contamination, treat glassware and water with a solution of DMDC diluted 1/1000 in distilled water for 24 h.
2. DMDC is destroyed by autoclaving at 121 °C for 30 min.
3. Immerge all plastic devices (e. g., electrophoresis tanks, combs, etc.) for 1 h in a 2% Absolve solution in water and rinse with DMDC-treated water.
4. All materials should be manipulated with gloves subsequent to treatment.

Protocol 2   Preparation of the bacterial cells

1. Culture conditions should be strictly maintained to avoid variations, other than those that are to be examined. Care must be taken to control both growth conditions and treatment of the cells. When possible, use the same batch of medium. Also, keep temperature, aeration, and agitation conditions identical for both cultures.
2. When using strains of different genetic backgrounds, it is better to collect cells at the same optical densities even if the growth rates are different. Treat cultures identically after growth. In particular, avoid putting a culture in ice to wait for the other one to reach the right optical density. Centrifuge cells under exactly the same conditions.
3. Poor cell lysis may result in low RNA yields. Optimize lysis by using conditions that are best adapted to the organism under study.

Protocol 3   RNA preparation

Warning!!! Manipulators must wear gloves during all the RNA preparation and manipulation steps.

Protocols without use of phenol

1. The High Pure RNA Purification kit (Roche Diagnostics GmbH) gives good results with several gram-positive and gram-negative bacteria. The kit can be used as described by the manufacturer, with one slight modification, described below.
2. Follow the manufacturer's protocol exactly for one round of purification.
3. To minimize the quantity of contaminating DNA, deposit the eluted RNA (100 µl) onto a new column provided in the kit after addition of the Lysis/Binding buffer (step 3). Then follow the protocol until the last step.
   Generally, we performed the final elution step using 70 µl of elution buffer. 22 µl of the RNA solution is put aside to perform quality control. The rest of the RNA solution can be used directly in first strand synthesis.
   If the quantity of RNA obtained is insufficient, it is possible to use several columns on which samples of the same cellular lysate are applied. Eluates are then combined. If needed, precipitate RNA (see below) to reduce the final volume.

Protocols with phenol extraction

Several protocols have been described using phenol. We present herein details of a protocol used for isolation of total RNA from *E. coli*.

1. After growing cells in the selected medium, mix 7 ml of the cell culture with 7 ml of absolute ethanol (100%) and centrifuge for 15 min at 4500 rpm.
2. Resuspend the pellet in 300 µl of solution 1.
3. Then add 600 µl of hot acid phenol (60 °C), followed by 100 µl of water.
4. Incubate the sample at 60 °C for 2 min, and mix several times by tube inversion.

5. After centrifugation, remove and conserve the aqueous phase and re-extract with hot phenol as described above.

6. Remove and conserve the aqueous phase solution and precipitate the RNA. Other protocols describing isolation of total RNA from *E. coli, S. cerevisiae,* and *B. subtilis* are available (http://www.microarrays.org/protocols.html) [4].

RNA obtained with either protocol can be solubilized in elution buffer or in water and conserved in aliquots at –20 °C for several weeks. For longer conservation, store the RNA at –80 °C in the ethanol solution used for precipitation (see below).

Protocol 4   RNA precipitation

1. To one volume of RNA, add 1/10 volume of 3 M sodium acetate and 2.5 volumes of absolute ethanol.
2. Place the tube for 30 min at –80 °C, then centrifuge 30 min at 10,000 rpm.
3. Resuspend the pellet with 1 ml of 70% ethanol.
4. Centrifuge 10 min at 10,000 rpm.
5. Air-dry the pellet.
6. Resuspend in the desired volume of DMPC-treated water (usually 75 µl when starting with 10 ml of culture), and incubate 10 min at 65 °C to completely dissolve the RNA sample.

Protocol 5   RNA quality control

Electrophoresis on agarose gels in TBE buffer

1. Load 1 to 2 µl of the RNA solution on a mini 1% agarose gel in 0.5 X TBE buffer.
2. After migration, stain the gel with ethidium bromide (in water) and analyze under UV light to verify the integrity of the 16S and 23S ribosomal RNA bands (see Fig. 2).

Quantification and purity of the RNA preparation

1. Dilute 5 to 10 µl of the RNA solution in 1 ml of distilled water and measure the absorbance of the solution at 260 nm using a spectrophotometer.
2. Calculate the RNA concentration applying the following correspondence: 1 OD unit at 260 nm is equivalent to a 40 µg/ml RNA solution.
3. Purity of the RNA preparation also can be tested by measuring the ratio of ODs at 260 and 280 nm. A pure RNA sample gives a ratio of between 1.8 and 2.1.

Checking for DNA contamination

Using two primers that amplify a DNA fragment from the genomic DNA you are working with, perform PCR with the RNA preparation as template. We generally perform the PCR test on a 0.5 to 1 µg RNA sample and do the same on a chromosomal sample (2 ng) as a positive control. A negative control is also performed in the absence of any genomic DNA. The presence of an amplification product indicates that contaminating DNA is present in the sample.

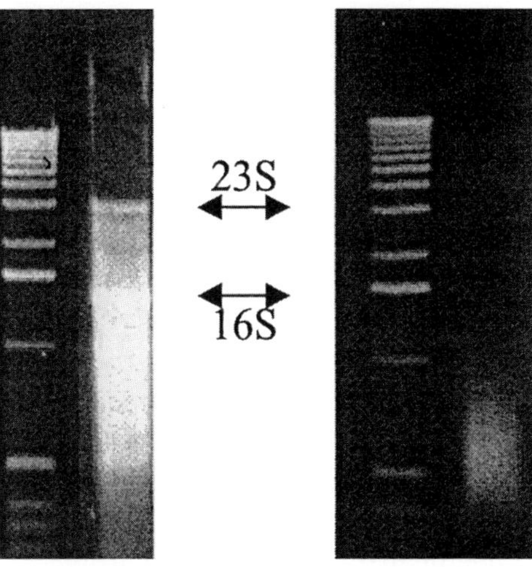

Figure 2 RNA quality control by gel electrophoresis. The gel on the left shows an RNA preparation that can be readily used to prepare cDNA. In contrast, the RNA sample run on the gel on the right is degraded and not suitable for transcriptome analyses. Positions of 23S and 16 S rRNA are indicated by arrows.

Protocol 6   cDNA synthesis

1. Mix RNA (between 5 and 10 µg) with 3 µg of random hexameric primers in a final volume of 15 µl.
2. Incubate at 70 °C for 3 min, then put in ice for 1 min.
3. Add the following cold mix: 1 µl dATP 10 mM, 1 µl dGTP 10 mM, 1 µl dTTP 10 mM, 6 µl 5X superscript-buffer, and 3 µl DTT 0.1 M.
4. Incubate 5 min at 25 °C before adding 1.5 µl of superscript II at 200 U/µl (Gibco-Brl) and 2 µl dCTP [α^{33}P] at 10 µCi/µl.
5. Incubate 5 min at 25 °C, followed by 1 h at 42 °C and 15 min at 70 °C.

Specific primers can be used instead of random primers. In this case, refer to the protocol provided by primer suppliers. However, we point out that random primers give better results than specific primers [5].

Protocol 7   cDNA cleanup

cDNA can be purified using the Qiaquick PCR purification kit (Qiagen) following the supplier's protocol. Alternatively, use Microcon-30, which in our hands gives the best yields.

Microcon purification procedure

1. Add water to the RT mix to obtain a final volume of 450 µl, and then load the sample on the Microcon filter already connected to its collection tube.
2. Centrifuge the assembled filter and tube at 12,000 rpm for 8 min.
3. Discard the flow-through and apply 400 µl of water on the Microcon filter.
4. Spin again at 12,000 rpm for 4 min.
5. Repeat this washing step with water two more times.
6. Invert the Microcon filter and place it on a new collection tube.
7. To recover the cDNA, centrifuge 3 min at 14,000 rpm.

Protocol 8   Hybridization

1. DNA macroarrays, either homemade or manufactured, very often consist of large (around 10 cm × 20 cm), positively charged nylon membranes. Rinse membranes in 50 ml 2x SSPE (diluted solution 2) for 5 min.
2. Before starting the prehybridization step, pre-warm the hybridization oven and the hybridization solution (solution 3) at 65 °C.
3. Pre-hybridize the membrane using 5 ml of warm hybridization solution in a roller bottle (6 rpm) for 1 h at 65 °C.
4. The entire labeled cDNA, prepared as described above, is added to 3 ml of hybridization solution.
5. Denature the sample for 10 min at 90–95 °C, and add it to the membrane after having decanted and discarded the prehybridization solution.
6. Hybridize the blot for 15 h at 65 °C. After this step, the hybridization solution can either be saved for future use or discarded appropriately.
7. Wash the blot 3 times in 50 ml washing solution (solution 4) for 5 min at room temperature and 3 times in 50 ml of pre-warmed washing solution at 65 °C for 20 min in the hybridization oven (6 rpm).
8. Remove the blot from the roller bottle and lay it on a sheet of blotting paper for a few minutes, taking care not to let the array dry completely (stripping will be less efficient!).
9. Wrap the blot in clear plastic food wrap, and expose it to a phosphorimager screen.
10. Check that no wrinkles are present in the clear plastic wrap separating the screen from the array.

The blot can be stripped (see below) and rehybridized up to four times.

Protocol 9   Stripping the arrays

1. Immerse the array in the stripping solution (solution 5) that had first been boiled, and incubate in a microwave oven for 20 min.
2. Wrap the array in plastic food wrap and expose to a phosphoimager screen.
3. Analyze the image, and compare it with the image obtained before stripping. If significant signals persist, perform the stripping protocol again using fresh stripping solution.
4. If not used for a new hybridization, the stripped arrays are wrapped in plastic wrap and then stocked at –20 °C.

Protocol 10   Exposure to phosphorimager screens

Several phosphorimager brands and their corresponding screens are commercially available (e. g., the Storm series of scanners from Amersham Biosciences or the Bas series from Fujifilm). For greater spot resolution, imaging screens should be scanned at a 50 µm rather than a 100 µm pixel size. The file sizes of 50 µm scans may be extremely large. It is advisable to crop the images after scanning to one field per image file. Typically, 1 to 3 days exposure to a phosphor imager screen will yield quantifiable results. It may be necessary to perform several exposures for different times to be able to quantify signals in cases of strong differences in expression levels.

Protocol 11   Image analysis

Once digitalized, images obtained for both control and test experiments are analyzed for quantification. Different image analysis software programs are commercially available; there is free access of some of these programs on the Web. ArrayVision software (Imaging Research, Inc.) is often recommended for macroarray analysis. For a recent comparison of the capabilities of the more frequently used programs, see http://dea-genomique.univ-rennes1.fr/ (in French).

4   Troubleshooting

Problems	Reasons	Solutions
RNA is degraded	RNA degradation occurred during agarose gel migration	Use anti-RNase products to treat all parts of the apparatus. Use RNase-free agarose and buffers
	RNA degradation occurred during the extraction procedure	Ensure that reagents, tips and tubes are RNase-free. Wear gloves at all times
RNA is contaminated with DNA	Treatment with DNase was not efficient	Test efficiency of DNase and treat once with the enzyme
Image is blurry	Improper contact between screen and blot	Make sure that the plastic wrap is flat against the blot and that the blot cannot shift within the cassette
Presence of non-uniform dark area on the image	Poor washing	
No signal or very low signal	Inefficient probe labeling	Check labeling efficiency of the probe
Unable to strip the blot	The blot dried at some point	If you want to reuse your blot, do not allow it to dry at any step

General considerations

While α^{32}P (half-life, 14.3 days) emits more energy and has a higher range of radiation than α^{33}P (half life, 25.4 days), it is strongly advised to use the latter for labeling. Sharper signals are produced by α^{33}P, which avoids diffusion and allows easier quantification of adjacent spots.

Gene expression signals are quantified from phosphor-imager-generated image files. However, results of arrays can be recorded as autoradiographs; this method provides pictures with sharper edged boundaries and offers a qualitative "back-up" to phosphorimaging files. A two-fold difference in expression levels is generally considered significant. However, single spot data are subject to variability. As such, experiments should be repeated at least three times, preferably with different RNA preparations, to minimize misclassification rates and to assure significance of the results [6]. The use of glass slides with fluorescent probes allows simultaneous hybridization with two samples; in this system the test and control samples are examined simultaneously on the same slide, unlike macroarrays where the test and control samples are hybridized on separate membranes. Special care should be taken for the normalization of the arrays. When arrays contain numerous spots, signals can be normalized from different arrays by representing the "average" spot signal as a percentage of the total signal from all spots on the array. Alternatively, the signal can be normalized as a percentage of positive controls such as genomic DNA spots. Membranes can be reused several times. However, dehybridization alters the arrays and subsequent hybridization to the spotted DNA. Membrane quality can be evaluated by the form of spots and by reproducibility between duplicated spots (if present). We recommend that the same stripping and recycling procedures be used for membranes that are to be directly compared.

5   Applications

The major application of prokaryotic macroarrays is for the characterization of transcriptional profiles. It provides a quick means to identify putative targets of a given regulator or to assess transcriptional modifications that are due to a change in growth conditions. Arrays can be very useful for the characterization of genes with unknown function: modification of the transcriptional pattern as a result of the inactivation or overproduction of a given gene can reveal which cellular processes the particular gene is involved in or how its expression is modified under the given condition.

Genomic arrays are also used for purposes other than transcriptome profiles. Gene transfer or loss among related strains can be traced by examining the distribution of its DNA using chromosomal DNA as probe on macroarrays. For example, the evolutionary history of the entire *E. coli* chromosome among different strains was examined by this method [7]. Changes in DNA copy number also can be evaluated to study replication fork movement [8].

Genome-wide DNA-binding protein interactions can be studied by array technologies. DNA fragments bound to a given protein can be isolated and then hybridized against genomic arrays. This technique was used, for example, to locate or identify DNA-binding sites of transcription factors [9]. DNA specific to different organisms also can be spotted on membranes. Resulting arrays can then be used for microbiological characterization of complex bacterial samples such as ecosystems.

6   Remarks and conclusions

Transcriptome analyses rely on two main techniques, membranes with radio-labeled probes and glass slides with fluorescent-labeled probes. At present, nylon arrays with α^{33}P-labeled radioactive probes provide the most sensitive and reliable results [10] and are easier to introduce and handle in an academic laboratory than glass slides hybridized with fluorescent-labeled probes. Indeed, when a phosphor screen reader is available, the use of this technique requires equipment routinely found in molecular biology laboratories. The field of high-density DNA arrays is undergoing rapid developments. For example the term microarray, which previously referred exclusively to the use of glass slides, is now also used to describe recently developed small-sized filters [10]. Nylon supports, once used only in conjunction with radioactive probes, are now analyzed with colorimetric probes [11]. It is likely that alternative procedures will soon be proposed, and researchers should closely follow progress in the field.

Acknowledgments

We are grateful to Sandy Gruss for reading of the manuscript. Transcriptome analysis in our laboratories is supported by grants from "Programme de Recherche Fondamentale en Microbiologie, Maladies Infectieuses et Parasitaires", from Hoechst Marion Roussel and "Puces à ADN" from CNRS.

Further reading

Supplement to Nature Genetics vol 21 Jan 1999, p 1–60
Arraybase link page: http://www.sghms.ac.uk/depts/medmicro/bugs/ARRAYBA-SE.htm
DNA microarray (genome chip): http://www.gene-chips.com/
Microarray: http://ihome.cuhk.edu.hk/~b400559/array.html
Microarray project: http://www.nhgri.nih.gov/DIR/Microarray/main.html
Nylon microarrays: http://tagc.univ-mrs.fr/microarrays/
The Stanford microarray database: http://www.dnachip.org/index.shtml
The web resource for gene expression and DNA microarray technologies: http://industry.ebi.ac.uk/~alan/MicroArray/

References

1   Lockhart DJ, Winzeler EA (2000) Genomics, gene expression and DNA arrays. *Nature* 405: 827–836

2   DeRisi JL, Iyer VR, Brown PO (1997) Exploring the metabolic and genetic control of gene expression on a genomic scale. *Science* 278: 680–686

3   Tao H, Bausch C, Richmond C et al. (1999) Functional genomics: expression analysis of *Escherichia coli* growing on minimal and rich media. *J Bacteriol* 181: 6425–6440

4   Sekowska A, Robin S, Daudin JJ et al. (2001) Extracting biological information from DNA arrays: an unexpected link between arginine and methionine metabolism in *Bacillus subtilis*. *Genome Biol* 2: research0019.1–0019.2

5   Arfin SM, Long AD, Ito ET et al. (2000) Global gene expression profiling in *Escherichia coli* K12. The effects of integration host factor. *J Biol Chem* 275: 29672–29684

6   Lee ML, Kuo FC, Whitmore GA, Sklar J (2000) Importance of replication in microarray gene expression studies: statistical methods and evidence from repetitive cDNA hybridizations. *Proc Natl Acad Sci U S A* 97: 9834–9839

7   Ochman H, Jones IB (2000) Evolutionary dynamics of full genome content in *Escherichia coli*. *EMBO J* 19: 6637–6643

8   Khodursky AB, Peter BJ, Schmid MB et al. (2000) Analysis of topoisomerase function in bacterial replication fork movement: use of DNA microarrays. *Proc Natl Acad Sci U S A* 97: 9419–9424

9   Ren B, Robert F, Wyrick JJ et al. (2000) Genome-wide location and function of DNA binding proteins. *Science* 290: 2306–2309

10 Bertucci F, Bernard K, Loriod B et al. (1999) Sensitivity issues in DNA array-based expression measurements and performance of nylon microarrays for small samples. *Hum Mol Genet* 8: 1715–1722

11 Chen JJ, Wu R, Yang PC et al. (1998) Profiling expression patterns and isolating differentially expressed genes by cDNA microarray system with colorimetry detection. *Genomics* 51: 313–324

13 Prokaryotic Proteomics

Cécile Lelong and Thierry Rabilloud

Contents

Methods and Tools in Biosciences and Medicine
Prokaryotic Genomics, ed. by M. Blot
© 2003 Birkhäuser Verlag Basel/Switzerland

1   Introduction

Proteomic methods, and particularly 2D-gel electrophoresis, the large-scale analysis of proteins, have become one of the main ways to understand gene function in the post-genomic era. Here, we will focus on three microorganisms: *Escherichia coli*, *Bacillus subtilis*, and *Synechocystis sp.* PCC 6803, all of which are studied as model organisms. *E. coli* is the gram-negative bacteria model, *B. subtilis*, the gram-positive, and *Synechocystis*, the oxygen photosynthesis bacteria model. As a consequence of their model status, they are all totally sequenced, which is essential for proteomics [1–4].

The first part of this chapter will be dedicated to the protein extraction methods, which must achieve several goals [5]:

1) Be easily reproducible: for this purpose, chemical methods are preferable and very efficient with exponential growth cultures. Specific methods have been developed for each type of organism. But we will also describe physical methods (sonication, French press, or glass beads) that can be useful in the case of difficulties to break the cells, e. g., in studies using late stationary growth cultures.
2) Break macromolecular interactions in order to yield separate polypeptide chains.
3) Prevent any artifactual modification of the polypeptides in the solubilisation medium.
4) Allow the easy removal of substances that may interfere with 2-D electro-phoresis.
5) Keep proteins in solutions during the 2-D electrophoresis process.

One of the main requirements for 2-D protocols is reproducibility of spot position. When an adequate method of protein extraction is adopted, the technique of isoelectric focusing on immobilized pH gradients (IPGs) is ideally suited to provided highly reproducible IEF separations, which represent the first step of 2D electrophoresis. Sections 3.4 and 3.5 will describe the second dimension procedure and two different staining protocols. In the last para-graph, an example of 2D study will be presented involving the comparison of three strains: a wild-type strain and two experimentally evolved strains (2000 and 5000 generations) [6].

2 Materials

2.1 Equipment

- Multiphor II (Amersham Pharmacia, Sweden)
- Protean II® (Biorad)

2.2 Solutions, reagents, and buffers

Protein extractions and solubilization
- Anti-protease solutions, such as PMSF, leupeptin, pepstatin, or benzamidine, may be added to TE and SDT buffer to prevent proteolysis.
- TE: 10 mM Tris-HCl pH 7.5, 1 mM EDTA
- Phosphate buffer: mix 7 g Na_2HPO_4, 3 g $NaHPO_4$, and 4 g NaCl in 1 l of distilled water.
- SDT: 0.05 M Trizma Base, 0.2 M DTT, 0.3% SDS, and 1 mM EDTA.
- DR: 1 mg/ml DNAse, 0.25 mg/ml RNAse, 0.05 M Tris-HCL pH 7.5, and 0.05 M $MgCl_2$.
- Lysis buffer: 9.9M urea, 4% NP40, 0.1M DTT and 2.2% ampholine.
- The three solutions, SDT, DR, and lysis buffer, are filtered, aliquoted (1 ml, 50µl and 1 ml, respectively), and then frozen at –20 °C.

First dimension
- Rehydration solution
- Prepare the rehydration stock solution by mixing, for 12 ml of concentrated (1.25x) solution, 2.3 g thiourea (2.5 M), 6.3 g urea (8.7 M), 0.6 g CHAPS (2% w/v), 150 µl Ampholytes (1% v/v) in 4 ml of warm distilled water ($\approx$ 35 °C). Stock solution can be stored at –20 °C.
- For (1x) rehydration solution: add 50 µl 1 M DTT to 950 µl distilled water and 4 ml 1.25x stock solution.

Second dimension
Equilibration solution
- Mix 25 ml Tris (120 g/l-HCl 0.8 M), 25 ml of 20% SDS, 100 ml of 60% glycerol and 72 g urea. Divide the equilibration solution in two parts of 100 ml each: in the first one, add 0.8 g DTT and in the second one, add 4 g iodoacetamide.
- Gel solution for one gel (10% total acrylamide): Mix 10 ml of 110 g/l Tris/0.6 M HCl, 20 ml of 30% Acrylamide:Bis-acrylamide, 60 ml of distilled water, and 20 µl of TEMED. Add 400 µl of 10% ammonium persulfate and immediately pour the gel. Overlay the gel with 1 ml 2-butanol until complete polymerisation. Remove the 2-butanol before migration or storage of the gel.
- Gels may be stored in wet atmosphere: Place the gel in a box containing wet paper filter, pipet distilled water on the top of the gel, and close the box.

Migration solutions
- Agarose solution: mix by heating 1 g of low melting point agarose, 12.5 ml of Tris 120 g/l HCl 0.8 M, 2 ml of 20% SDS, 2 ml of bromophenol blue in 100 ml H_2O total volume. The solution is aliquoted and stored at 4 °C.
- Cathode buffer: 6 g/l Tris, 25 g/l Taurine and 1 g/l SDS.
- Anode buffer: 6 g/l Tris, 28 g/l Glycine, 1 g/l SDS.

Staining solutions
Silver staining
- Sodium thiosulfate solution: 10% (w/v) solution of crystalline sodium thiosulfate pentahydrate in water. Small volumes of this solution are prepared fresh every week and stored at room temperature.
- Formaldehyde: formaldehyde stands for commercial 37–40% formaldehyde. This is stable for months at room temperature.
- Ethanol: A technical ethanol grade of alcohol can be used and 95% ethanol can be used instead of absolute ethanol, without any volume correction. However, denatured ethanol cannot be used.
- Silver nitrate solution: 1 N silver nitrate solution is less expensive than solid silver nitrate and is stable for months if kept in a dark cold place.
- Fixation solution 1: 30% Ethanol, 10% acetic acid in distilled water.
- Fixation solution 2: 30% Ethanol, 5% acetic acid in distilled water.
- Sensitization solution: 2 ml 10% sodium thiosulfate solution in 1 l distilled water
- Silver nitrate solution: 12.5 ml silver nitrate solution (1 N) in 1 l distilled water
- Development solution: 35 g potassium carbonate, 125 µl 10% sodium thiosulfate solution, 300 µl of formaldehyde in 1 l distilled water.
- Stop solution: 40 g Tris, 20 ml acetic acid in 1 l of distilled water.

Coomassie staining
- Coomassie solution: 480 mg Coomassie Blue, 100 ml ethanol, 32 ml acetic acid and distilled water to a final volume of 400 ml.

Note
Lysis buffer and rehydration buffer contain urea and should not be heated above 37 °C. On heating, urea breaks down, and the resulting modifications cause carbamylation of proteins, resulting in streaking in the isoelectric focusing dimension.

3 Methods

3.1 Protein extractions and solubilization

Protein extraction methods are specific for each organism. They depend on chemical characteristics of each bacterial species. For example, *Bacillus subtilis*, which is gram positive, needs a specific procedure to disrupt the cell wall. For *Synechocystis*, a photosynthetic organism, thylakoid membranes and associated pigments have to be removed from soluble protein to obtain an adequate spot separation. Specific treatments are required to solubilize proteins of the thylakoid membrane and membrane proteins in a general manner [7]. This will not be treated here. We will focus only on the extraction of soluble proteins. This section is composed of two parts: the first one is dedicated to "physical" methods of protein extraction and the second one to "chemical" methods. Physical methods are more adapted for difficult protein extractions, such as extraction of proteins from stationary phase growth culture, but are less reproducible than chemical methods. However, physical methods can be applied to all organisms, with some adaptations.

Protocol 1   Growth culture

1. Liquid cultures are stopped at an OD_{600nm} of 0.6–0.8 for *E. coli* and *B. subtilis* and OD_{580nm} of 1–2, for *Synechocystis* 6803 by spin at 5,000 g (5000 rpm) for 10 min at 4 °C.
2. The pellets are washed twice with phosphate buffer (final volume 1.5 ml in microfuge): Transfer the culture resuspended in phosphate buffer to a microfuge tube, spin at g (13,000 rpm in microfuge) for 5 min at 4 °C, and remove the supernatant.
3. Then either freeze the samples at –20 °C or make the protein extracts.

Physical methods

Protocol 2   Sonication

1. Resuspend the cell pellet in cold TE (1 ml TE for 10 ml pelleted in microfuge tube) or lysis buffer (1 ml lysis buffer for 10 ml liquid culture pelleted in the microfuge tube).
2. The cell suspensions are sonicated using a microtip: 5–10 bursts (5 s each) using the lowest power setting.
3. Intensity, number, and time of the bursts have to be adjusted for each organism.
4. After sonication, the cell debris is removed by ultracentrifugation (60 min, 90,000 g).

5. The supernatant is directly used for the rehydration step after protein quantification or stored at –20 °C.
6. Samples in TE cannot exceed 1/5 of the total strip rehydration volume, while samples in lysis buffer can be used at any dilution.

Protocol 3   French press

1. Resuspend the cell pellets in 1 ml TE containing 20 µl DR solution.
2. The suspensions are passed four times through a French press pressure cell at 20,000 psi.
3. The homogenates are ultracentricentrifuged for 1 h at 200,000 x g.
4. Protein pellets are resuspended in 500–1000 µl lysis buffer (vortex vigorously the mixture to obtain homogenated solution). Protein samples can be stored at –20 °C until 2-D electrophoresis experiments.

Protocol 4   Glass beads

1. Resuspend the cell pellets in lysis buffer containing 1–3 g of glass beads (425–600 mm diameter; Sigma).
2. Vortex vigorously for 5 min.
3. Remove the cell debris by centrifugation (15 min at 10,000 x g) and use the supernatant as the soluble fraction buffer (vortex vigorously the mixture to obtain homogenated solution).
4. Protein samples can be stored at –20 °C until the 2-D electrophoresis experiments.

Chemical methods

Protocol 5   *Escherichia coli*

1. Resuspend cell pellets in SDT solution such that the protein concentration is 5–20 µg/µl, place the extract at 100 °C for 5 min, and then cool it on ice.
2. Add DR solution and incubate on ice for 10 min.
3. Finally, add the lysis buffer and vortex vigorously the mixture to obtain a homogeneous solution.
4. For 10 ml of culture at OD_{600nm} of 0.8, used volumes are 200 µl of SDT, 20 µl of DR and 800 µl of lysis buffer (vortex vigorously the mixture to obtain a homogeneous solution). Protein samples can be stored at –20 °C until the 2-D electrophoresis experiments.

Protocol 6   *Bacillus subtilis*

1. Resuspend cell pellets in 1 ml TE containing 10 µg/ml of lysozyme; incubate 10 min at 37 °C.

2. Add 100 μl of a solution containing 3% SDS and 50% 2-mercaptoethanol. Place the mixture at 100 °C for 5 min, then cool on ice.

3. Add 20 μl of DR solution and incubate on ice for 10 min.

4. Cell debris is removed by centrifugation at 10,000 g (13,000 rpm in microfuge tubes) for 20 min at 4 °C.

5. The supernatant is diluted in an equal volume of acetone, and incubated at least 20 min on ice.

6. Protein pellets are obtained by centrifugation of the mix at 10,000 x g (13,000 rpm in microfuge tubes) for 30 min at 4 °C. They are dried and resuspended in 400–1000 μl lysis buffer (vortex vigorously the mixture to obtain homogenated solution). Protein samples can be stored at –20 °C until the 2-D electrophoresis experiments.

Protocol 7   *Synechocystis* 6803

1. The harvested cyanobacterial cells are resuspended in 500 μl of chloroform containing 0.07% 2-mercaptoethanol.

2. The cell suspensions are incubated five times alternatively, in liquid N_2 for 5 min and in water bath at 37 °C for 5 min.

3. Add 500 μl acetone containing 0.07% 2-mercaptoethanol to the mixture, and then incubated at least 1 h at –20 °C.

4. Protein pellets are obtained by centrifugation at 10,000 g (13,000 rpm in microfuge tubes) for 45 min at 4 °C.

5. The pellets are washed with 500 μl acetone containing 0.07% 2-mercaptoethanol and dried.

6. Protein pellets are resuspended in 500–1000 μl lysis buffer (vortex vigorously the mixture to obtain a homogeneous solution).

7. Protein samples can be stored at –20 °C until 2-D electrophoresis experiments.

3.2   Protein quantitation

Protein concentrations of each preparation are estimated using Bradford assay (Biorad kit) on a protein extract diluted to 1/10. At this dilution, an adequate buffer blank can compensate for the interference caused by SDS.

3.3   First dimension

Gel electrophoresis in the first dimension is carried out using immobilized pH gradient gels (pH 4–7/18 cm or pH 3–10/18 cm) with a horizontal apparatus (Multiphor II; Amersham Pharmacia).

Rehydration step

Dry strips (Immobiline Drystrips or laboratory made) are rehydrated with the protein sample (100–500 µg, depending on sample and gel size) added to the rehydration solution. The optimal rehydration solution depends on the sample. Typically, it contains urea, a nonionic or zwitterionic detergent, ampholytes matching the pH range of the IPG strip, a reducing agent (DTT), and a dye. Urea solubilizes and denatures proteins, unfolding them by disruption of noncovalent interactions and exposing internal amino acids. Using thiourea in addition to urea increases the chaotropic power of the sample solution. Denaturation by urea/thiourea induced the exposure of the totality of the protein's hydrophobic residues to the solvent. This increases the potential for hydrophobic interactions, especially when lipids are present in the sample. This explains why detergents are almost always included in the urea-based solubilization mixtures for 2-D electrophoresis. Detergents solubilize lipids and hydrophobic proteins and minimize protein aggregation. The detergent must have zero net charge: use only nonionic and zwitterionic detergents such as CHAPS, Triton X-100, or NP-40. Ionic detergents, as SDS, used for the initial solubilization prior to IEF can be tolerated because they are in low amounts and provided that high concentrations of urea and nonionic or zwitterionic detergents are present to ensure complete removal of the SDS from the proteins during IEF.

Ampholytes can improve separations and sample solubility, particularly with high sample loads. Reducing agents cleaves disulfide bridges to allow proteins to unfold completely. A tracking dye (bromophenol blue or orange G) provides a monitor for the progress of the IEF at the beginning of the migration.

Protocol 8   Rehydration in a reswelling tray

1. Mix protein extract (100–500 µg) and rehydration solution (1x) to a final volume of 300–350 µl (for a 3 mm-wide and 18 cm long strip).
2. Pipet 3 µl of dye solution (bromophenol blue) into each slot of the reswelling tray.
3. Add each mix (protein + rehydration solution) into each slot. Remove any large bubbles. To ensure complete protein extract uptake, do not apply excess rehydration solution.
4. Position the IPG strip (with gel side facing down) and lower the strip into the mix.
5. Overlay each IPG strip with 1 ml of mineral oil to minimize evaporation and urea crystallization.

Rehydration needs at least 16 h at room temperature (overnight).

Isofocalization

The IEF is performed in multiphor II system from Pharmacia Amersham. IEF is conducted at very high voltage and very low currents. It must be noted that some high-voltage power supplies have securities that prevent them from delivering very low intensities. Thus, dedicated power supplies must be used.

A typical protocol proceeds through a series of voltage steps: low initial voltage minimizes sample aggregation and allows the parallel separation of proteins with differing salt concentrations. Many factors affect time and steps required for complete isofocalization, e. g., sample and rehydration solution composition, strip length, pH gradient. Empirical determination is required for optimal results. However, a total volt-hours comprised between 60,000 and 70,000 is recommended for 18 cm strips. IEF also needs efficient cooling for precise temperature control.

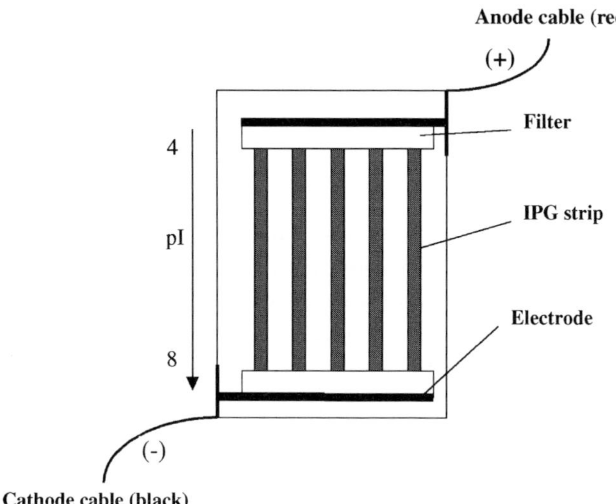

Figure 1 Scheme of the rehydrated IPG strips in Multiphor II apparatus

Protocol 9 Prepare IPG strip for isofocalization

1. Remove the rehydrated IPG strips from the reswelling tray and place them onto the cooling plate of an electrofocusing chamber: the acidic part of the IPG is positioned close to the anode cable (red) and the basic part close to the cathode cable (black).
2. Wet two sheets of filter paper with water. Place a piece of wet filter paper on each extremity of the IPG strips. Press each electrode on the wet sheet of filter.
3. Cover the entire surface with mineral oil.
4. Check the temperature to 20 °C and apply current. Table 1 gives an example of protocol for 18 cm IPG strip, pH 4–8.
5. At the end of the IEF, the strips can be stored at −20 °C or directly used for the second dimension.

Table 1 Isofocalization protocol for 18 cm strip, pH 4–8

Length	pH range	step	Voltage (V)	Current (mA)	Power (W)	Vh
18 cm	4–8	1	100	10	10	1
		2	100	10	10	100
		3	300	10	10	100
		4	300	10	10	300
		5	3450	10	10	2000
		6	3450	10	10	60,000–70,000

3.4 Second dimension

Equilibration

Before the second dimension, the strips have to be equilibrated in a buffer adapted to the migration in the second dimension. The equilibration buffer contains Tris buffer, SDS, glycerol, urea, and reductant in a first step. In the second step, the reductant is replaced by iodoacetamide to alkylate thiol groups on proteins and prevent their reoxidation during migration.

Rocking in equilibration solution containing DTT (10 ml/18 cm strip) first incubates each strip. After 15 min, the same volume of the equilibration buffer containing iodoacetamide replaces the solution. Incubate at least 15 min.

At the end of the equilibration, the strips can be stored at –20 °C or directly used for the migration.

Migration

Remove the strips from the equilibration buffer and put at the top of the gel. The strips are embedded onto SDS/PAGE gel with agarose solution. Gels are run at 10 °C at 25 V for 1 h then 12.5 W/gel for 4–5 h.

Note: Cooling the second-dimension gels is essential. The best way is to connect slab gel tanks can to a recirculating chiller. The optimum temperature is the coolest possible without precipitation of the SDS. Practically, this is between 10 °C and 20 °C

3.5 Staining

Batches of gels (up to 6 gel/box) can be stained. For a batch of three to four medium-size gels (e. g., 160 × 200 × x1.5 mm), 1 l of the required solution is used. Five hundred milliliters of solution is used for one or two gels. Batch processing can be used for every step longer than 5 min, except for image development, where 1 gel/box is mandatory. For steps shorter than 5 min, the gels should be dipped individually in the corresponding reagents. When gels must be handled individually, they are manipulated with gloved hands. Except for development or short steps, where occasional hand agitation of the staining vessel is convenient, constant agitation is required for all steps. A reciprocal ("flip-flop") shaker is used at 30–40 strokes/min.

Silver staining

The main advantage of the silver staining method is its high sensibility, 100 times higher than staining with Coomassie blue. Silver staining protocol comprises four major steps. The first step is fixation, which aims at insolubilizing the proteins in the gels and removing the interfering compounds present in the 2-D gels. The second step is sensitization which aims at increasing the subsequent image formation. The third step is silver impregnation. The fourth and last step is image development.

Protocol 10   Silver staining

1. At the end of migration, fix the gels in fixation solution 1 at least 30 min.
2. Replace fixation solution 1 by fixation solution 2 and soak the gels overnight.
3. Rinse in 20% ethanol 10 min.
4. Rinse in 10% ethanol 10 min.
5. Rinse 2 × 10 min in distilled water.
6. To sensitize, soak the gels for 1 min (one gel at time) in sensitisation solution.
7. Rinse 2 × 1 min in distilled water.
8. Impregnate for at least 30 min in silver nitrate solution.
9. Rinse in water 15 sec.
10. Develop image (10–20 min) in development solution.
11. Stop development (30–60 min) in stop solution.
12. Rinse with water (several changes) prior to drying or densitometry.

Coomassie blue staining

The Coomassie method should be used mainly to extract and characterise proteins in gels. There must be 500 µg–1 mg of protein extracts loaded during the rehydration step to visualise spots by Coomassie blue staining.

Protocol 11   Coomassie blue staining

1. At the end of migration, soak the gels in Coomassie blue solution at least 1 h.
2. Change 30% ethanol several times until maximal contrast (i. e., visible spots on a transparent background) is achieved.

3.6 Protein characterization

Unequivocal protein characterization has been achieved by several methods, including amino acid analysis and N-terminal or internal peptide sequencing by Edman chemistry. Nowadays, these methods have been completely superseded by mass spectrometry-based methods, using either peptide mass fingerprinting or tandem mass spectrometry-based methods (MS/MS) [8, 9]. The detailed description of these methods is beyond the scope of this chapter, as very special and expensive instrumentation is required. However, it can be mentioned that peptide mass fingerprinting is very efficient in terms of cost and throughput for completely sequenced organisms. Otherwise, it is often safer to use MS/MS methods, which are more expensive, more difficult, and provide a lesser throughput.

4 Applications
Comparison of an ancestor *Escherichia coli* B strain (wild type) and two experimentally evolved (2000 and 5000 generations) strains

2D-gel electrophoresis can be used for many purposes. In all cases the main goal is the comparison of the protein expression patterns between two populations of bacteria. These populations should be: (1) the same strain grown in different conditions, e. g., different carbon sources, one before and one after a (salt, cold, heat...etc) stress; (2) a wild-type strain and a mutant strain deleted for one (or several) known gene(s) to study the consequences of such mutations; or (3) two isolates belonging to the same species to identify differences between them.

In this part, we will describe proteomic experiments that allow comparison of three *E. coli* clones: an ancestor strain and two isolated from the ancestor after 2000 and 5000 generations [6] of experimental evolution in a constant environment [10]. The three strains are described Chapter 3. Genetic diversity in these experimental evolution populations of *E. coli* has been followed using IS elements as genotypic markers. Mapping IS insertions gave precious information about the target of natural selection [6, 11]. In particular, one of them has been found to be beneficial for the evolved strain [12]. Using IS fingerprinting, some genetic modifications resulting from the evolution have been highlighted. The analysis of some of the same evolved strains was performed by 2-D gel. Additional differences, not detected with IS fingerprints, were discovered.

Overnight culture of each strain (wild type and 2000 and 5000 generations strains), in M9 minimal medium, was diluted to an OD_{600nm} of 0.05 in fresh media. Cultures were grown at 37 °C by shaking until OD_{600nm} of 0.6 was reached. Protein extracts were obtained as described in Section 3.1 using the chemical protocol specific of *E. coli*.

IPG of 18 cm, pH 4–8, was rehydrated overnight with 400 µl of IEF solution that contained 150 µg of each protein extract. Isoelectric focusing was conducted for 65,000 V.h using Multiphor II system (Amersham-Pharmacia) at 20 °C.

For the second dimension, the IPG was equilibrated for 15 min by rocking, in equilibration solution containing first DTT and then iodoacetamide during the same time (see section 3.3). The strips were then embedded onto SDS/PAGE gel with agarose solution. Gels were run at 10 °C at 25 V for 1 h then 12.5 W/gel for 5 h at 10 °C.

Finally, gels were stained using silver nitrate protocol section 3.5.

Gels were performed three times for each strain using two independent cultures.

Protein characterisation: Using the same proteins extracts, 2D-gels were performed with 500 µg of proteins. Gels were stained using Coomassie blue protocol. Spots of interest were cut in gel and submitted to mass spectrometry experiments. Identification is in progress.

Protein expression patterns of the evolved strains (2000 and 5000 strains) were compared with those of the ancestor strain (WT). Fourteen proteins showed a different pattern: nine spots have disappeared and five have a decreased expression in both evolved strains (Fig. 2).

Using mass spectrometry technology, two of them have already been identified: they corresponded to the PykF and the RbsB proteins (Fig. 2 and Tab. 2). Their disappearance in the evolved strains was correlated with genetic results: IS fingerprints have shown that *pykF* gene and *rbsB* gene were both inactivated by an IS*150* insertion and a deletion, respectively [11, 12]. The identification of the other spots are in progress, using mass spectrometry methodology. Some genetic experiments will be conducted to understand the phenotype (disappearence or appearence) of some them. (Complete results will be submitted later on).

Table 2   Proteins identified by mass spectrometry

Protein name	Accession number	MS method	Theoretical PI/PM	Observed pI/PM	Phenotype	Function
PykF	P14178	MS	5.68/50.3	5.9/55	Disappeared	Pyruvate kinase
RbsB	P02925	MS	7.26/30.95	6.3/30	Disappeared	D-ribose-binding periplasmic protein precursor

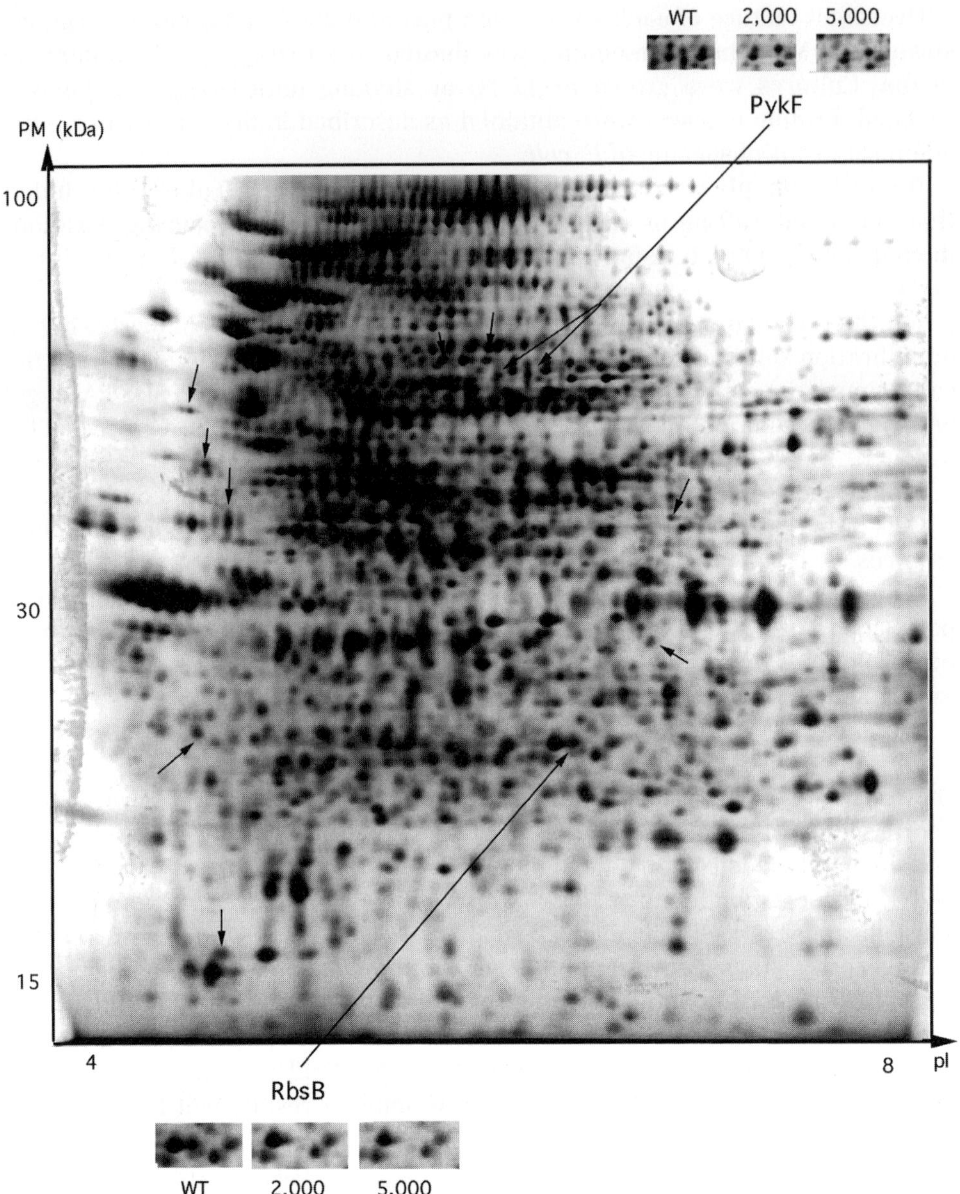

Figure 2 2-D gel electrophoresis of the wild-type strain.
Arrows point out spots, which have shown phenotypes (disappearance or decreasing) in the evolved strains by comparison with the wild-type strain. (*)The three boxes show spot phenotypes in the three strains (wild type [WT], 2000 and 5000 generations evolved strains) for the already-identified spots.

In conclusion, 2D-gel experiments have allowed discrimination and characterization of strains that belong to the same species but are genetically different. They have given a reproducible fingerprint of protein expression in a given growth condition that may be used to identify a strain. They have also given new trails to better understand not only genetic organization of such strains but also their physiological adaptation to a given environment. The latest characteristics of proteomics experiments also could be applied to all physiological studies.

5   Troubleshooting

Information on troubleshooting is mentionned in the text where it fits to the methodological context

References

1   Blattner FR, Plunkett G III, Bloch CA et al. (1997) The complete genome sequence of *Escherichia coli* K-12. *Science* 277: 1453–74

2   Kaneko T, Sato S, Kotani H et al. (1996) Sequence analysis of the genome of the unicellular Cyanobacterium *Synechocystis* sp. strain PCC6803. II. Sequence determination of the entire genome and assignment of potential protein-coding regions *DNA Res* 3: 109–136

3   Kaneko T, Sato S, Kotani H et al. (1996) Sequence analysis of the genome of the unicellular Cyanobacterium *Synechocystis* sp. strain PCC6803. II. Sequence determination of the entire genome and assignment of potential protein-coding regions (supplement) *DNA Res* 3: 185–209

4   Kunst F, Ogasawara N, Moszer I et al. (1997) The complete genome sequence of the Gram-positive bacterium *Bacillus subtilis*. *Nature* 390: 249–256

5   Rabilloud T (1996) Solubilization of proteins for electrophoretic analyses. *Electrophoresis* 17: 813–829

6   Papadopoulos D, Schneider D, Meier-Eiss J et al. (1999) Genomic evolution during a 10,000-generation experiment with bacteria. *PNAS* 96: 3807–3812

7   Santoni V, Molloy M, Rabilloud T (2000) Membrane proteins and proteomics: un amour impossible? *Electrophoresis* 21: 1054–1070

8   Griffin TJ, Aebersold R (2000) Advances in proteom analyses by mass spectrometry. *J Biol Chem* 276: 45497–45500

9   Aebersold R, Goodlett DR (2001) Mass spectrometry in proteomics. *Chem Rev* 101: 269–295

10   Lenski RE, Rose MR, Simpson SC et al. (1991) Long-term experimental evolution in *Escherichia coli*. I. Adaptation and divergence during 2000 generations. *Am Nat* 138: 1315–1341

11   Schneider D, Duperchy E, Coursange E, et al. (2000) Long-term experimental evolution in *Escherichia coli*. IX. Characterisation of insertion sequence-mediated mutations and rearrangements. *Genetics* 156: 477–488

12   Cooper VS, Schneider D, Blot M et al. (2001) Mechanisms causing rapid and parallel losses of ribose catabolism in evolving populations of *Escherichia coli* B. *J Bacteriol* 183: 2834–2841

14 Intein-mediated Protein Purification

Shaorong Chong and Francine B. Perler

Contents

Methods and Tools in Biosciences and Medicine
Prokaryotic Genomics, ed. by M. Blot
© 2003 Birkhäuser Verlag Basel/Switzerland

1   Introduction

Protein-splicing elements are remarkable multifunctional protein domains that mediate their post-translational excision from a precursor protein. The intervening sequence (called the intein) plus the first carboxy-terminal amino acid of the surrounding host protein (called the extein), direct cleavage of the peptide bonds on both sides of the intein and the formation of a native peptide bond between the two extein fragments. The majority of inteins also encode a homing endonuclease subdomain that initiates mobilization of the intein gene into an extein gene homolog that does not contain the intein [1]. Understanding the mechanism of protein splicing has led to a variety of intein-mediated applications. This chapter describes the use of modified inteins in protein purification

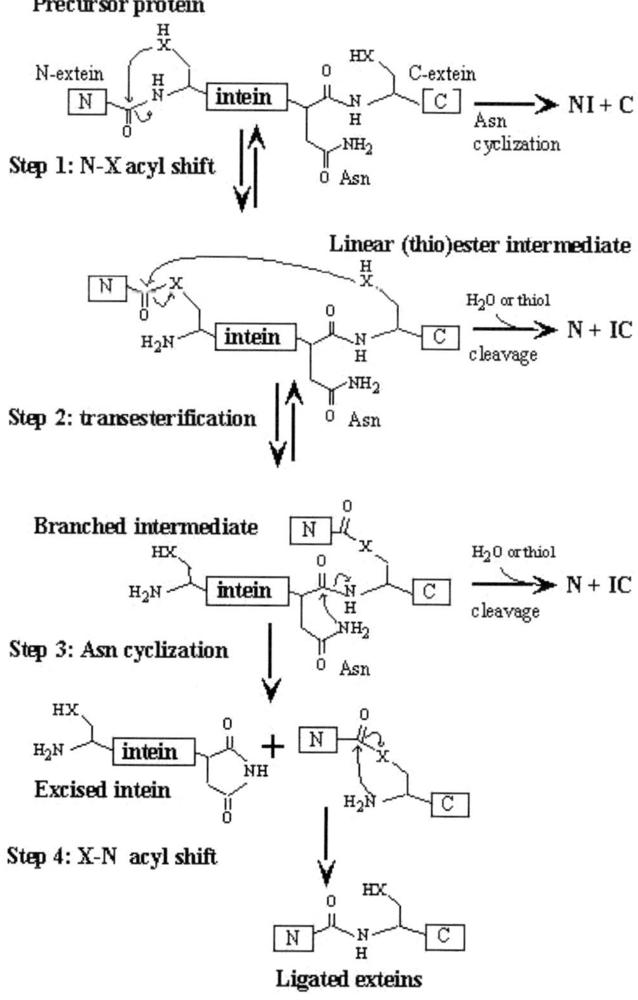

Figure 1   The chemical mechanism of protein splicing. The protein splicing mechanism is depicted with X representing the oxygen or sulfur atom of Ser, Thr, or Cys; N representing the amino-extein; and C representing the carboxy-extein. Intein-based protein purification vectors utilize controlled, single splice site cleavage reactions.

and the special chemical reactivities of proteins isolated after thiol cleavage when the carboxy-terminus of the target protein is fused to an intein vector.

The mechanism of protein splicing (Fig. 1) was rapidly determined soon after the discovery of conditions permitting controllable splicing of a purified precursor [2] and has been extensively reviewed (see Further Reading). It consists of a four-step pathway involving (1) activation of the amino-terminal splice junction, (2) cleavage of the upstream splice junction combined with ligation of the two extein fragments, (3) removal of the intein after cleavage of the peptide bond at the downstream splice junction, and (4) an acyl rearrangement to yield a peptide bond between the extein fragments. Although each step is a simple, classic chemical reaction facilitated by the intein, protein splicing requires an exquisitely programmed coordination of these steps. In the absence of this coordination, single splice junction cleavage leads to dead-end side products (Fig. 1). Hydrolysis or thiol-mediated cleavage of the (thio)ester in the linear or branched intermediate results in cleavage at the upstream splice junction. The cleavage side reaction at the downstream splice junction occurs when Asn cyclization precedes the other steps in the protein splicing pathway.

A new class of protein purification vectors has been designed based on the ability of inteins to cleave the peptide bond at a single splice junction [3–10]. These vectors use modified inteins that permit controlled cleavage to separate the protein of interest from the intein (Figs. 1–3). The intein, in turn, is fused to

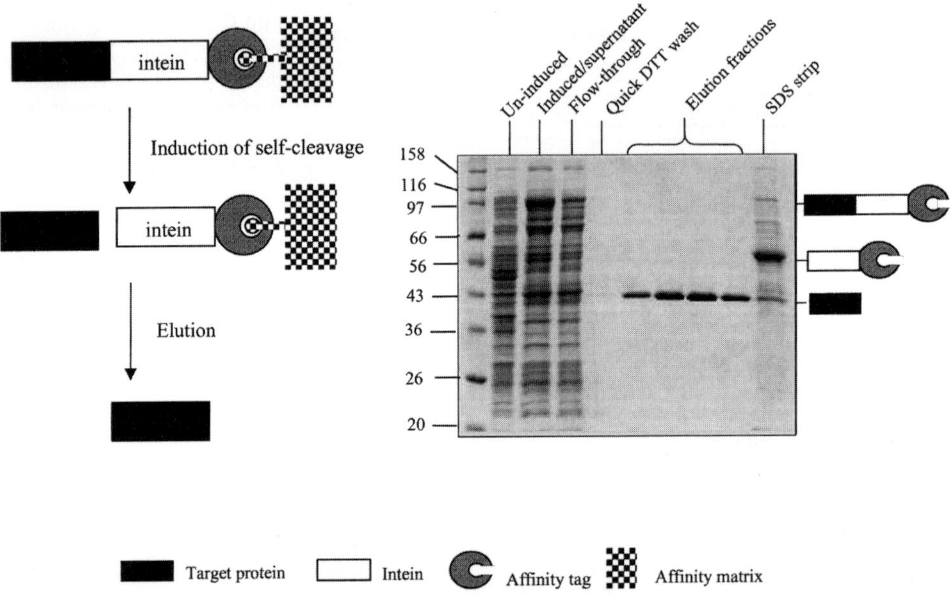

Figure 2 Illustration of intein-mediated purification of recombinant HhaI methyltransferase. The cell extract is passed over a chitin column (checkered box) allowing the fusion protein to bind using the chitin affinity tag. After washing away unbound material, intein cleavage is initiated with DTT. Free target protein (black box) is eluted after incubation with DTT. Stripping the column with SDS reveals almost complete cleavage with only minor amounts of precursor compared to the intein::CBD fragment. Coomassie blue stained SDS-PAGE gel. The size of the molecular weight markers (kDa) is listed on the left side of the gel.

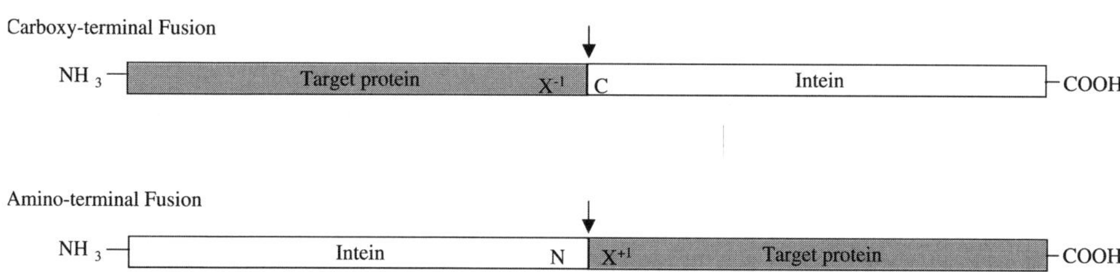

Figure 3 Fusion of a target protein to an intein. The carboxy-terminal fusion: the carboxy-terminus of a target protein is fused to the amino-terminal Cys (C) of an intein; the last residue of the target protein is called the minus 1 residue (X^{-1}). *The amino-terminal fusion:* the amino-terminus of a target protein is fused to the carboxy-terminal Asn (N) of an intein; the first residue of the target protein is called the plus 1 residue (X^{+1}). "X" represents any amino acid.

an affinity tag for precursor purification. A chitin-binding domain (CBD) is most frequently used as the affinity tag because of its small size (5 kDa), high affinity, and broad buffer compatibility. The precursor protein containing the target protein is first purified away from the other soluble *Escherichia coli* proteins using this affinity tag to bind the precursor to a solid support. The target protein is then cleaved away from the remainder of the fusion protein while still attached to the affinity matrix. Washing the column thus results in release of the purified target protein after a single combined purification and cleavage chromatography step (Fig. 2).

Intein-mediated protein purification has many advantages. The target protein can be purified free of tags or extra amino acids after a single affinity purification step. Because no expensive proteases are required to free the target protein from the affinity tag, cleavage within the target protein is not a problem. In some instances, expression in intein vectors yielded more soluble protein or protein with higher specific activity than when expressed with other protein purification systems [11, 12].

2 Materials

2.1 Media

LB broth (Solution 1 per liter), 10 g tryptone (Difco, Detroit, MI); 5 g yeast extract (Difco, Detroit, MI); 10 g NaCl. Adjust the pH to 7.0 with NaOH.

2.2 *E. coli* strains

Cloning should be conducted in a non-expression host, especially if the protein of interest is toxic or could contribute to plasmid instability. Cloning strains: *E. coli* ER2267, DH5, JM109 (New England Biolabs, Beverly, MA). Expression strains when a T7 promoter is used: ER2566 (New England Biolabs, Beverly, MA), BL21 (DE3) (or other derivatives) (Novagen, Madison, WI) or BL21-CodonPlus(DE3)-RIL (Stratagene, La Jolla, CA). Expression strains when a *lac* or *tac* promoter is used: ER2267, JM109, or TB1 (New England Biolabs, Beverly, MA).

2.3 Solutions, reagents, and buffers

Cell Lysis Buffer (Solution 2): 20 mM Na-HEPES or Tris-HCl or Na-Phosphate (pH 6.0–9.0 depending on application (see section 3.4), 500 mM NaCl (usable range: 50–1000 mM NaCl), 1 mM EDTA (optional), 0.1% Triton X-100 or 0.2% Tween 20 (optional), 20 μM phenylmethylsulfonyl fluoride (PMSF) (as protease inhibitor, optional), 1 mM TCEP [tris-(2-carboxyethyl)phosphine] or TCCP [tris-(2-cyanoethyl)phosphine] (for proteins sensitive to oxidation, optional).

Column Buffer (Solution 3): 20 mM Na-HEPES or Tris-HCl or Na-Phosphate (pH 6.0–9.0 depending on application), 500 mM NaCl (usable range: 50–1000 mM NaCl), 1 mM EDTA (optional). When used for the induction of on-column cleavage, the Column Buffer also contains 50 mM dithiothreitol (DTT), β-mercaptoethanol, free cysteine, or 2-mercaptoethanesulfonic acid (MESNA).

Stripping Buffer (Solution 4): 20 mM Na-HEPES or Tris-HCl; pH 8.0; 500 mM NaCl; 1% SDS (kept at room temperature).

SDS-PAGE Sample Buffer (Solution 5): 70 mM Tris-HCl, pH 6.8; 33 mM NaCl; 1 mM NaEDTA; 2% SDS; 40 mM DTT; 0.01% bromphenol blue; 40% glycerol.

Chitin beads (Solution 6): Available from New England Biolabs (Beverly, MA). Binding capacity: ~ 4 mg protein per l ml (bed volume) chitin beads. Chitin beads are supplied as a slurry in 20% ethanol and should be stored 4 °C.

Anti-CBD antibody (Solution 7): Rabbit serum raised against the *Bacillus circulans* chitin binding domain (CBD), available from New England Biolabs (Beverly, MA) for Western blot analysis.

Expression vectors: all intein vectors (carboxy-terminal vectors: pCYB series, pTYB 1 to 4, pTXB 1 and 3, pTWIN 1 and 2, and pTEG series; amino-terminal vectors: pTYB 11 and 12, pBSC 1 and 3, pTWIN 1 and 2) are from New England Biolabs (Beverly, MA).

3   Methods

3.1   Choosing an intein fusion vector

There are two classes of the intein fusion vectors (Fig. 3): (1) the carboxy-terminal fusion vectors – the target protein is fused upstream of the intein, i. e., the carboxy-terminus of the target protein is fused to the amino-terminal cysteine residue of the intein and (2) the amino-terminal fusion vectors – the target protein is fused downstream of the intein, i. e., the amino-terminus of the target protein is fused to the carboxy-terminal asparagine of the intein. In a carboxy-terminal fusion vector, the carboxy-terminal residue of the target protein is called "minus 1 residue (X^{-1})", and in an amino-terminal fusion vector, the amino-terminal residue of the target protein is called "plus 1 residue (X^{+1})" (Fig. 3).

The carboxy-terminal fusion vectors (Tab. 1) include: pCYB series (*lac* promoter), pTYB 1 to 4, pTXB1 and 3, pTWIN 1, and pTEG series. All except pCYB use a T7 promoter. The amino-terminal vectors (Tab. 2) include: pTYB 11 and 12, pBSC 1 and 3, pTWIN 1 and 2, all of which use a T7 promoter. pTWIN 1 and 2 can be used as both carboxy- and amino-terminal vectors depending on the cloning sites employed.

Table 1   Effect of the carboxy-terminal residue of the target protein (the minus 1 residue) on intein-mediated cleavage in carboxy-terminal fusion vectors

Cloning vector	Induction of cleavage	Effect of the target protein carboxy-terminal residue on cleavage efficiency (%)	
pCYB series, pTYB1–4	Addition of thiols	**4 °C**	
		Gly, Ala	>90%
		Met, Phe, Tyr, Trp, Gln, Lys, Ser, Thr, Glu, His	>50%
		Leu, Val, Ile	<50%
		Arg, Asp	*In vivo* cleavage
		Asn, Cys, Pro	<5%
		>16 °C	
		Gly, Ala, Met, Phe, Tyr, Trp, Gln, Lys, Ser, Thr, Glu, His, Leu, Val, Ile	>50%
		Arg, Asp	In vivo cleavage
		Asn, Cys, Pro	<5%

Cloning vector	Induction of cleavage	Effect of the target protein carboxy-terminal residue on cleavage efficiency (%)	
		4 °C	
		Gly, Ala, Gln	>80%
		Tyr, Leu, Met, Ser, His, Phe	>50%
		Ile, Asn, Cys, Thr, Lys, Arg	<50%
		Trp, Glu, Pro	<20%
pTEG series	Addition of thiols	Asp, Val	*In vivo* cleavage
		>16 °C	
		Gly, Ala, Gln, Tyr, Leu, Met, Ser, His, Phe, Ile, Asn, Cys, Thr, Lys, Arg	>80%
		Trp, Glu, Pro	>20%
		Asp, Val	*In vivo* cleavage
		23 °C	
pTXB 1 and 3, pTWIN	Addition of thiols	Gly, Ala, Gln, Tyr, Leu, Met, His, Phe, Ile, Asn, Cys, Thr, Lys, Arg, Trp	>50%
		Ser, Glu, Pro	<50%
		Asp	*In vivo* cleavage

Table 2   Effect of the amino-terminal residue of the target protein (the plus 1 residue) on intein-mediated cleavage in amino-terminal fusion vectors

Cloning vector	Induction of cleavage	Effect of the target protein amino-terminal residue on cleavage efficiency (%)	
		4 °C	
		Met, Ala, Gln	>60%
		Gly, Leu, Asn, Trp, Phe, Tyr	>40%
		Val, Ile, Asp, Glu, Lys, Arg, His, Thr, Pro	< 20%
		Ser, Cys	*In vivo* splicing
pTYB11 and 12	Addition of thiols	**>16 °C**	
		Met, Ala, Gln, Gly, Leu, Asn, Trp, Phe, Tyr	>90%
		Val, Ile, Asp, Glu, Lys, Arg, His, Thr	>70%
		Pro	<10%
		Ser, Cys	*In vivo* splicing
		>25 °C	
pBSC1 and 3	pH & temperature	Ser, Cys, Ala, His	>75%
		Gly, Met, Glu, Asp, Trp, Phe, Tyr, Val, Thr	10–75%
		Gln, Asn, Leu, Ile, Arg, Lys, Pro	<10%
pTWIN 1 and 2	pH & temperature	Same as pBSC 1 and 3	

The following factors affect the choice of the intein fusion vectors for optimal expression and purification of a recombinant protein: (1) sensitivity to reducing thiols such as DTT, (2) the amino-terminal or carboxy-terminal residue of the target protein that serves as the minus 1 or plus 1 residue in the intein fusion protein, and (3) extra residues that are added to the terminus of the target protein without affecting the structure and activity of the target protein.

To choose an appropriate intein fusion vector, follow the flow chart shown in Figure 4. If the target protein is sensitive to a high concentration of DTT (e. g., 50 mM), use a carboxy-terminal fusion vector and free cysteine to induce cleavage.

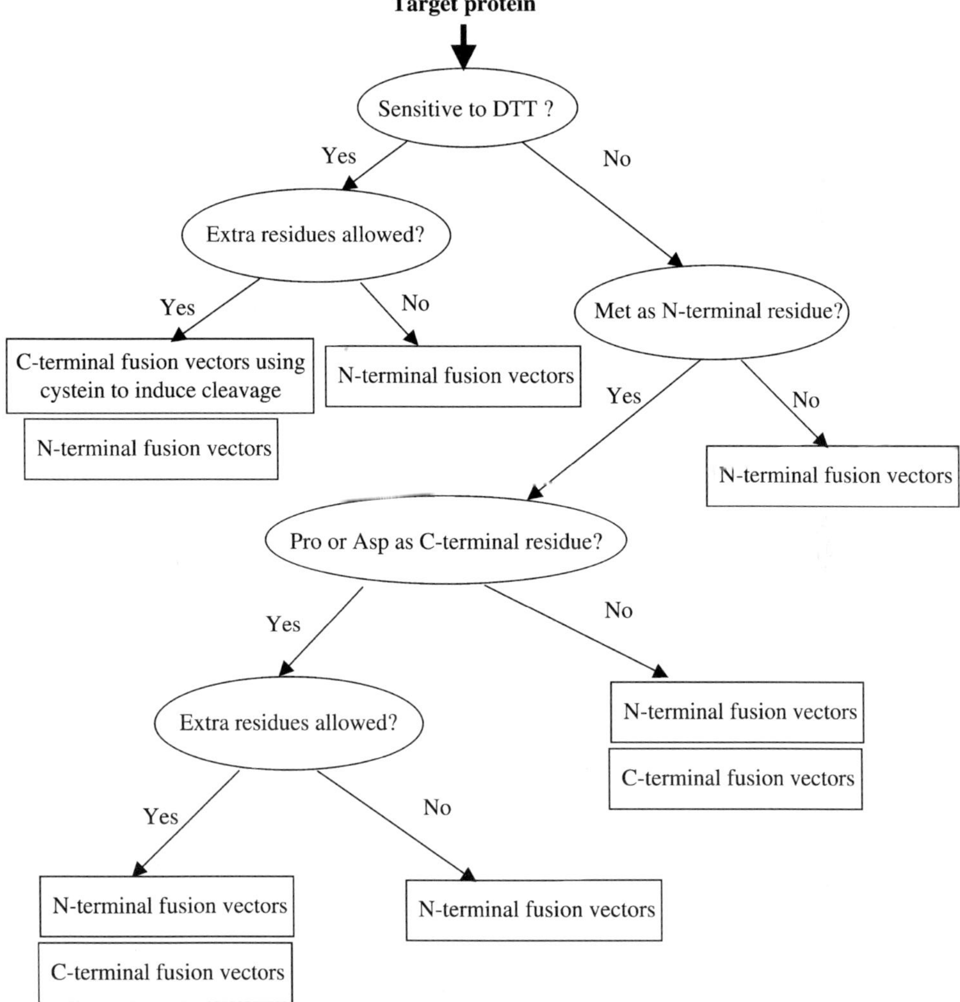

Figure 4   A flow chart for selecting an appropriate intein fusion vector. Several factors contribute to successful target protein purification, including sensitivity to thiols such as DTT, presence of extra residues, and the specific residue of the target protein that is to be cloned next to the intein. The variety of inteins in the various vectors allows some flexibility in finding a compatible vector for each target protein based on known preferences for proximal residues or on the method used to induce cleavage.

In this case, a cysteine residue will be covalently attached to the carboxy-terminus of the target protein. Alternatively, use a vector that induces cleavage by changing pH and/or temperature. An amino-terminal fusion vector allows a target protein to be purified without a Met as the amino-terminal residue. Certain minus 1 or plus 1 residues will inhibit intein cleavage (Tabs. 1 and 2). In general, Pro and Asp inhibit cleavage and cause > 90% *in vivo* cleavage, respectively. If a target protein contains Pro or Asp as the carboxy-terminal residue, use either an amino-terminal fusion vector or a carboxy-terminal fusion vector with an extra residue added to the carboxy-terminus of the target protein. If possible, one should try both carboxy- and amino-terminal fusions for each target protein to maximize the chance of successful expression and purification. Several amino- and carboxy-terminal intein vectors contain the same restriction enzyme sites in the polylinker to allow simple testing of the target protein in both contexts.

3.2   Cloning a target gene into an intein vector

Figure 5 illustrates cloning a target gene into an intein vector. If appropriate restriction enzyme sites are available to allow direct transfer of the target gene, go to Protocol 3.

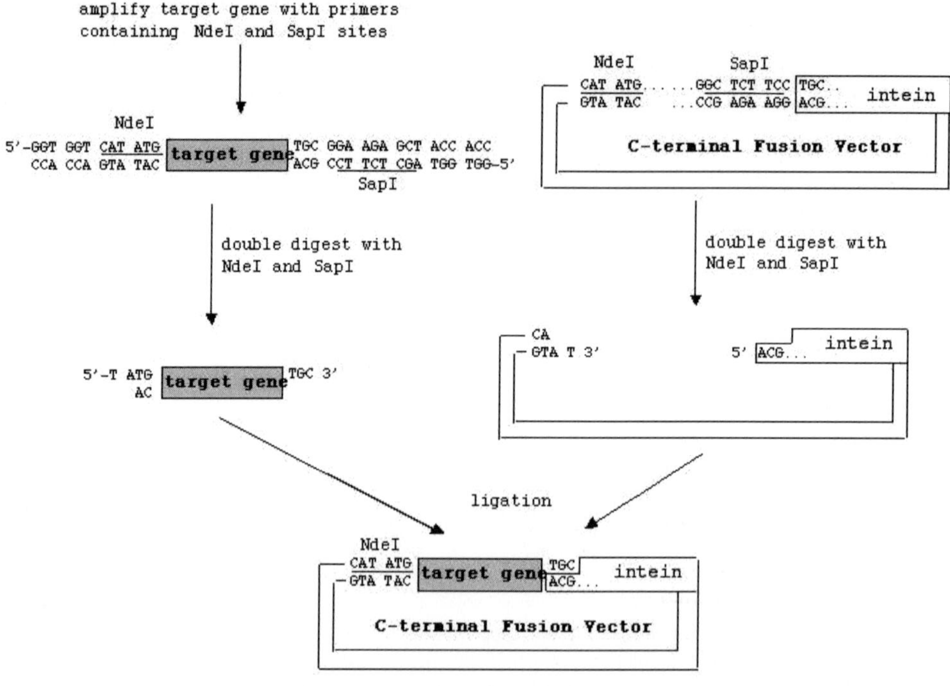

Figure 5A   Strategies for cloning target proteins. Schematic illustration of cloning an amplified target gene into (A) the *Nde*I and *Sap*I sites of a carboxy-terminal fusion vector or into (B) the *Sap*I and *Pst*I sites of an amino-terminal fusion vector.

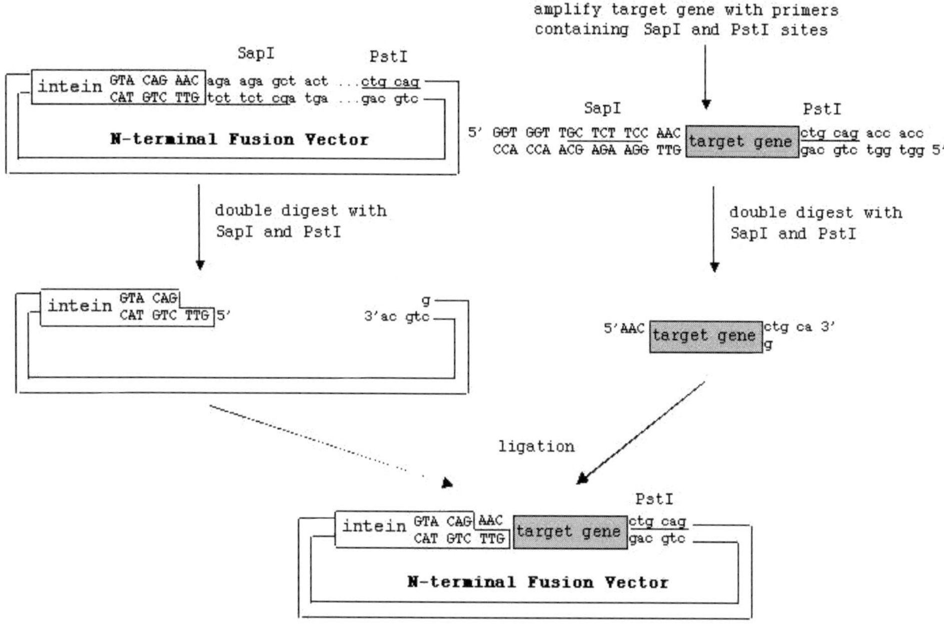

Figure 5B

Protocol 1   Primer design

Primers should contain appropriate restriction enzyme sites and extra nucleotides at the 5' end to ensure efficient restriction enzyme digest. Examples of primer sequences are shown in Table 3.

Table 3   Examples of primer design for cloning in carboxy-terminal and amino-terminal fusion vectors

Cloning vector	Cloning site	Primer sequence[a]
Carboxy-terminal fusion vectors	– *Nde*I (forward)[b]	5'-GGT GGT CAT ATG NNN NNN...-3'
	– *Sap*I (reverse)[c]	5'-GGT GGT TGC TCT TCC GCA NNN NNN...-3'
Amino-terminal fusion vectors	– *Sap*I (forward)[d]	5'-GGT GGT TGC TCT TCC AAC NNN NNN... -3'
	– *Pst*I (reverse)[e]	5'-GGT GGT CTG CAG TTA NNN NNN... -3'

[a]   The target gene starts at "5'-NNN NNN...". Restriction enzyme sites are underlined. Use of a "GGT GGT" sequence (or any other GC-rich 6-nucleotide sequence) at the 5' end of the primer is to ensure efficient DNA cleavage by the restriction enzyme when the restriction enzyme site is close to the 5' end.

[b]   *Nde*I contains the starting codon (ATG) of the target gene.

[c,d]   *Sap*I site is not regenerated after cloning.

[e]   The complement of the TAA (TTA) stop codon is incorporated in the reverse primer.

Protocol 2   Amplification of a target gene

A target gene is amplified using the above designed primers in a polymerase chain reaction (PCR). A typical PCR mixture (100 µl) contains 1x DNA polymerase buffer, 4 mM $MgSO_4$, 400 µM of each dNTP, 1 µM each of the forward primer and reverse primer, and 1.0 unit of DNA polymerase. The DNA polymerase, the DNA polymerase buffer, and other PCR reagents are available from Roche Molecular Biochemicals (Indianapolis, IN) or QIAGEN Inc. (Valencia, CA). Amplification is carried out using a thermal cycler (PerkinElmer, Wellesley, MA) at 95 °C for 1 min, 55 °C for 1 min, and 72 °C for 1 min for 25 cycles. The primer annealing temperature (segment 2) is dependent on the sequence of the primer. The extension time at 72 °C is dependent on the length of the target gene, a good rule of thumb is 1 min per kb. After the PCR reaction, the reaction mixture is loaded onto a 1% agarose gel (American Bioanalytical, Natick, MA), and the band corresponding to the target gene is cut out with a razor blade. The DNA in the gel slice is extracted in a 50 µl Tris solution (pH 8.0) with a QIAquick PCR purification kit (QIAGEN Inc., Valencia, CA).

Protocol 3   Restriction enzyme digestion

The target gene and an intein fusion vector are digested with appropriate restriction enzymes in a typical reaction mixture (80 µl) containing 8 µl 10x restriction enzyme buffer and 10 units of restriction enzymes (per µg of DNA) and are incubated at 37 °C for 3 h. The reaction mixtures are then loaded onto a 1% agarose gel (American Bioanalytical, Natick, MA) and the fragments corresponding to the digested target gene and the vector are isolated and purified using a QIAquick PCR purification kit (QIAGEN Inc., Valencia, CA).

Protocol 4   Ligation

A typical ligation mixture (100 µl) contains 10 µl 10x ligase buffer (New England Biolabs), purified DNA fragments (in 3-to-1 molar ratio of the target gene to the intein vector), and 1 µl T4 DNA ligase (New England Biolabs). The ligation reaction is conducted at 16 °C overnight. Alternatively, the Quick ligation kit (New England Biolabs) can be used, and the ligation reaction can be conducted at room temperature for 10 min.

Protocol 5   Transformation

The ligated product (15 µl) is transformed into a cloning strain such as ER2688 or JM109 (New England Biolabs).

Protocol 6   Screening for target gene inserts

Plasmid DNA is isolated using QIAprep mini-columns (QIAGEN Inc., Valencia, CA). The plasmids containing the target gene can be screened by restriction enzyme digestion. Note that when *Sap*I is used, the *Sap*I site is lost after ligation (Fig. 5) and a nearby restriction enzyme site should be used for insert screening. Alternatively, one can use PCR to screen for inserts directly from colonies or from plasmid DNA.

3.3   Cell culture and protein expression

The intein vector containing the target gene is transformed into an appropriate *E. coli* expression strain and the intein fusion protein is expressed. Recombinant protein expression is normally affected by the following factors: (1) *E. coli* strain, (2) cell culture conditions (e. g., temperature, aeration, cell density, etc.), and (3) protein expression induction conditions (e. g., temperature, duration, IPTG concentrations, etc.).

Protocol 7   Transformation into an expression strain

To express the intein fusion gene using a T7 promoter, the vector is transformed into ER2566 or BL21(DE3). If the target gene inhibits cell growth, BL21(DE3)-pLysS or BL21(DE3)pLysE should be used to reduce basal level expression. If the target gene contains multiple rare codons for *E. coli* expression, BL21-Codon-Plus(DE3)-RIL may help. To express the intein fusion gene under a *lac* or *tac* promoter, the vector is transformed into ER2267, JM109, TB1 or other similar host.

Protocol 8   Cell culture

LB medium (Solution 1, 1 liter) containing 100 µg/ml ampicillin is inoculated with freshly grown colonies. Inoculation with an overnight starter culture is not recommended because it may result in low expression. Protein expression is often improved by using freshly transformed cells. The culture is incubated on a rotary shaker at 37 °C to a culture density (OD_{600nm}) of 0.5–1.0, corresponding to mid- or late log phase.

Protocol 9   Induction

Protein expression is induced by adding IPTG to a final concentration of 0.3–0.5 mM and by incubating the culture in a rotary shaker at 20–25 °C overnight. Some proteins may require lower IPTG concentrations or even no IPTG for soluble expression. In general, lower induction temperatures (as low as 16 °C) result in higher soluble protein expression and higher intein cleavage efficiency after purification. If induction is performed at 37 °C, reduce the induction time to 1–3 h. Note that high induction temperature may cause protein aggregation and inclusion body formation.

3.4   Protein purification

Protocol 10   Cell harvest

The cells from the above culture are spun down at 5000 x g for 10 min at 4 °C. After discarding the supernatant, the cell pellet is resuspended for immediate protein purification or can be stored at or below –20 °C.

The following steps should be performed on ice or at 4 °C, unless otherwise stated.

Protocol 11   Preparation of chitin column

Chitin beads (Solution 6, 20 ml per one-liter culture) are equilibrated at 4 °C with 10 volumes of Column Buffer (Solution 3, for pBSC1 and 3, pTWIN 1 and 2 as amino-terminal fusion vectors, pH 8.5 only; for all other vectors, pH 6.0–9.0 can be used).

Protocol 12   Preparation of crude cell extract

The cell pellet from a one-liter culture is resuspended in 60 ml ice-cold Cell Lysis Buffer (Solution 2, for pBSC1 and 3, pTWIN 1 and 2 as amino-terminal fusion vectors, pH 8.5 only; for all other vectors, pH 6.0–9.0 can be used). The cells are broken by either sonication or French press. A low level of lysozyme (10–20 µg/ml) also can be used (incubate at 4 °C for 1 h) to break cells, though lysozyme is known to bind and digest chitin. An increase in viscosity indicates that the cells are broken. If the mixture becomes extremely viscous, the cell lysate should be diluted and/or treated with protease-free DNase (10 µg/ml) plus $MgCl_2$ (5 mM) before proceeding to the next step. The clarified cell extract (supernatant) is obtained by centrifugation at 20,000 x g for 30 min.

Protocol 13   Loading clarified cell extract on a chitin column

The clarified extract should be loaded onto the chitin column slowly (0.5–1 ml/ min) to allow maximum binding. Samples from the load and the flow-through should be collected for future analysis in order to estimate the binding efficiency of the fusion precursor on the chitin column.

Protocol 14   Washing the chitin column

Because of the high affinity of the CBD for chitin beads, a higher flow rate (up to 2 ml/min) and stringent conditions, such as high salt concentrations (up to 1 M NaCl) and nonionic detergents, can be used to reduce nonspecific binding of other *E.coli* proteins. In general, at least 10 bed volumes of the Column Buffer (Solution 3, for pBSC1 and 3, pTWIN 1 and 2 as amino-terminal fusion vectors, pH 8.5 only; for all other vectors, pH 6.0–9.0 can be used) are required to thoroughly wash the column.

Protocol 15   Induction of on-column cleavage

The target protein is released from the chitin column when the chitin-bound intein undergoes self-cleavage in the presence of thiols (such as DTT, β-mercaptoethanol, or free cysteine) or upon changing pH and temperature. For all carboxy-terminal fusion vectors and amino-terminal fusion vectors pTYB11 and 12, the induction of the on-column cleavage is initiated by quickly flushing the column with five bed volumes of Column Buffer (Solution 3, pH 8.5) containing 50 mM DTT, β-mercaptoethanol, or free cysteine. Note that use of free cysteine (in the case of a target protein sensitive to DTT) results in an extra cysteine residue covalently attached to the carboxy-terminus of the target protein in a carboxy-terminal fusion vector. In the case of pTYB11 and 12, the cysteine is not attached to the target protein but to a 15-residue peptide (1.6 kDa) co-eluted with the target protein. If the carboxy-terminal vectors are used to generate a reactive carboxy-terminal α-thioester for IPL or EPL applications (see section 5), the induction of the on-column cleavage should use Column Buffer (Solution 3, pH 8.5) containing 50 mM 2-mercaptoethanesulfonic acid (MESNA). For the amino-terminal fusion vectors pBSC1 and 3, and pTWIN 1 and 2, the induction of on-column cleavage is initiated by quickly flushing the column with five bed volumes of Column Buffer (Solution 3) at pH 7. The cleavage reaction is allowed to continue by stopping the column flow. Incubate the column at 4–23 °C for 16–40 h. For the amino-terminal fusion vectors, incubation at 23 °C for 16 h is recommended. For carboxy-terminal vectors, the incubation temperature and time can vary depending on the target protein and the minus 1 residue (Tabs 2 and 3).

Protocol 16   Elution of the target protein

After completion of the cleavage reaction, the target protein is eluted from the
column using Column Buffer (Solution 3) or any specific storage buffer. The
target protein is normally detected in the first few fractions (3 ml per fraction).
When DTT or β-mercaptoethanol is used to induce cleavage, the thiols should be
removed by dialysis, as the high concentration of the thiols may affect the
activity of the target protein and the thiols may remain attached to the carboxy-
terminus of the target protein through a hydrolyzable thioester bond. When the
amino-terminal vectors pTYB11 and 12 are used, a small peptide (1.6 kDa) is
also cleaved from the intein tag and co-elutes with the target protein. Because
of its small molecular weight, the cleaved peptide cannot be detected on a
regular SDS-PAGE gel and normally can be removed from the target protein by
dialysis.

To determine the cleavage efficiency, 200 μl of resin is removed from the column
after complete elution of the cleaved target protein and then mixed with 100 μl
SDS-PAGE Sample Buffer (Solution 5) to remove the bound proteins. After
boiling for 5 min, the resin is spun down. The supernatant (3–10 μl) is directly
used for SDS-PAGE analyses.

Protocol 17   Regeneration of chitin resin

The chitin resin can be regenerated 4–5 times by the following protocol: Wash
the column with 3 bed volumes of Stripping Buffer (Solution 4) and 3 bed
volumes of 0.3 M NaOH. Allow the resin to soak for 30 min in 0.3 M NaOH
followed by a wash with an additional 7 bed volumes of 0.3 M NaOH. Wash with
20 bed volumes of water followed by 5 bed volumes of Column Buffer (Solution
3). The resin can be stored at 4 °C in Column Buffer (Solution 3) or Column
Buffer containing 0.02% sodium azide for long-term storage.

4   Troubleshooting

Problems	Reasons	Solutions
Cloning		
No inserts	Partial digestion of vector	Optimize conditions of double digestion
	Used SapI to check the insert but SapI site is lost after ligation	Use a nearby site or colony PCR to check inserts
	Target gene is toxic	More stringent control of basal level expression; use different strains
Mutations in the target gene	PCR with DNA polymerase of low fidelity	Use DNA polymerase of high fidelity or fewer cycles
	Target gene is toxic	More stringent control of basal level expression; use different strains
Expression		
No or low expression	No IPTG added	Add IPTG (0.2–1.0 mM)
	Weak promoter	Use stronger promoter such as T7 promoter
	Mutations in the target gene	Re-sequence the target gene
	Target gene is toxic	More stringent control of basal level expression; use different strains
	Rare codon usage	Replace rare codons with *E. coli* codons or use BL21-CodonPlus strains
	Fusion protein misfolded and degraded	Lower induction temperature Try different strains
	Fusion protein is insoluble	Lower IPTG concentration Lower induction temperature Use N-terminal intein vector Change strains
Expressed protein with incorrect size	Mutations in the target gene	Re-sequencing the target gene
	In vivo cleavage by protease or intein	Use protease deficient strain or change intein vector

Problems	Reasons	Solutions
Purification		
Fusion protein not binding to column	Loading too fast	Load the column at < 1 ml/min
	Not enough resin	Use at least 10 ml chitin resin for 1 liter culture
None or low level of the target protein eluted	Low level of protein expression	See above
	Most fusion protein insoluble	Lower IPTG concentration Lower induction temperature Use N-terminal intein vector Change strains
	In vivo cleavage	Use different intein vectors Change or add extra residue(s) between the target gene and the intein Lower induction temperature
	Low cleavage efficiency	Increase incubation time Increase incubation temperature Change the intein fusion vector Change or add extra residue(s) between the target gene and the intein
	Cleaved target protein aggregated on column	Use higher salt concentration (1M) or non-ionic detergent
Contaminating proteins	Insufficient washing of column	Washing with > 10 bed volume
	Association of target protein with other proteins (e. g. chaperones)	Use higher salt concentration (1M) or non-ionic detergent
Eluted protein with incorrect size	Mutations in the target gene	See above
	No expression of the target protein	See above

5   Applications

Intein-based protein purification vectors have been used as both simple purification systems and for more complicated protein-engineering applications. Intein-based protein purification is rapid and gentle. For example, when a single protein fused to an intein tag was mixed with potential interacting proteins, the entire complex was purified after on-column intein cleavage [13]. The broad range of allowable buffer constituents for chitin affinity purification provides the flexibility of eluting in buffers that are most compatible with the target protein or the next process. Intein-mediated protein purification also has been successful with proteins secreted from *Streptococcus gordonii* [14], proteins expressed by baculovirus [15], circularly permuted proteins [16], and proteins with cyclized backbones [17–20]. Backbone cyclization stabilizes many expressed proteins. An intein vector with a green fluorescent protein tag has been used to identify constructs that provide maximum soluble expression by merely screening for fluorescence [21].

Protein purification is not always simple and no single system will work with all proteins. Therefore, using vectors with different inteins or positioning the target protein on different ends of the fusion protein may improve expression of difficult proteins. Each intein has its own set of substrate preferences (preferred residues at the cleavage site, e. g., Tabs. 1 and 2). In the pTWIN vectors, a multiple cloning site polylinker is present between two different inteins. Depending on the cloning sites used, the amino-terminal or carboxy-terminal intein can be deleted from the fusion, providing a variety of inteins for single splice junction cleavage. Folding of target proteins, especially in heterologous host cell types, can sometimes be improved by having a fusion partner at the amino or carboxy terminus.

Carboxy-terminal fusion vectors yield target proteins that are "activated" for ligation to numerous types of molecules as a result of the presence of a carboxy-terminal α-thioester [8, 20, 22–32]. In the most commonly performed application, called intein-mediated protein ligation (IPL) or expressed protein ligation (EPL), a second polypeptide is ligated to the target protein (Fig. 6). IPL requires a polypeptide with a carboxy-terminal α-thioester and a second polypeptide with an amino-terminal Cys for ligation of the two polypeptides. The process can be iterated to build up a protein from more than two segments, allowing a combination of synthetic and biosynthetic components. Amino-terminal intein vectors (other than pTYB11 and 12) can be used to generate polypeptides with the required amino-terminal Cys, although other methods are available [8, 20, 28–32]. When a target protein is inserted between the two inteins in the pTWIN vectors, the purified protein can be isolated with an amino-terminal Cys and an activated carboxy-terminal α-thioester, allowing either cyclization or polymerization [30].

IPL can be used to express cytotoxic proteins [33]; to label proteins with fluorescent tags [23]; to label protein segments for NMR [34]; to insert non-ribosomal moieties such as biosensors or unnatural amino acids [23, 25, 29, 32, 35]; to generate circularly permuted forms or proteins with cyclized backbones [20, 30], phosphorylate proteins [23], or amidate proteins [27]; or to perform combinatorial peptide shuffling.

A. Intein-mediated expressed protein ligation

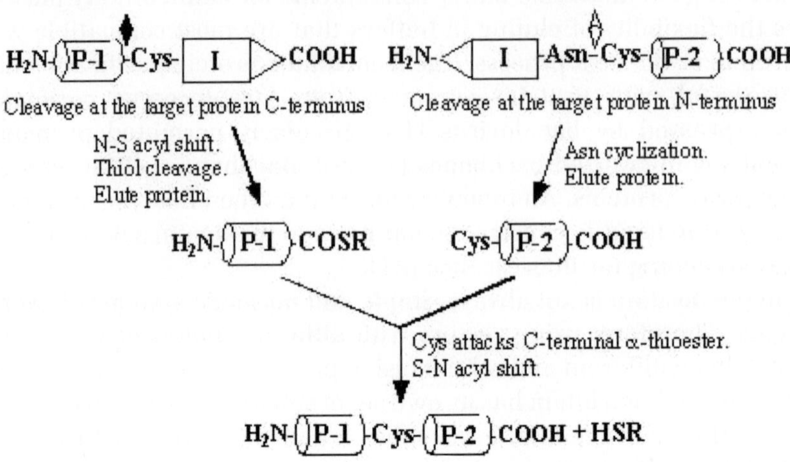

B. Cyclization or polymerization using pTWIN vectors

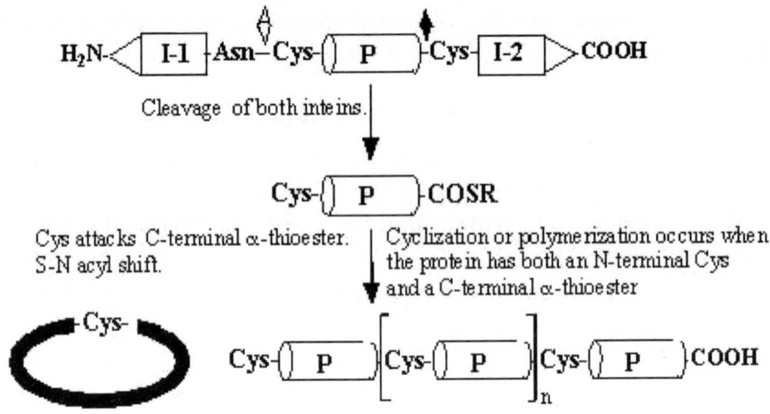

Figure 6 Intein-mediated protein ligation and cyclization. Schemes for IPL (A) and IPL-mediated cyclization and polymerization (B) are depicted. The carboxy-terminal α-thioester (COSR) group is generated by thiol -nduced cleavage of a target protein from a carboxy-terminal intein vector. A protein with an amino-terminal Cys can be generated by cleavage of a target protein from an amino-terminal intein vector or by other methods, such as proteolysis. In sequential IPL, multiple segments can be ligated by repeatedly adding masked polypeptides and cycling steps 1 through 3. Any moiety that is attached to a Cys or a peptide beginning with Cys can be added to a target protein using this technique. Any nucleophile capable of displacing the "R" group from the thioester also can be ligated to the carboxy-terminus of a target protein. Symbols: rectangle or I, intein; cylinder or P, target polypeptide; triangle, affinity purification tag, which can be adjacent to or within the intein; black diamond, amino-terminal splice junction cleavage; white diamond, carboxy-terminal splice junction cleavage; n, multiple units.

6   Remarks and conclusions

Inteins are one of the newest tools in the protein engineer's arsenal. Modified inteins provide flexible methods of purifying proteins as well as generating the building blocks for protein semisynthesis. This chapter has focused on commercially available intein-mediated purification vectors, but the noncommercially available vectors [5, 6, 9, 10, 21] work in very similar ways. The flow chart in Figure 4 provides a framework for avoiding known expression problems, such as incompatible terminal amino acids (see also Tabs. 1 and 2) or sensitivity to thiols. Several of the commercial vectors have compatible multiple cloning sites to permit ease of cloning target proteins into different vectors.

Further reading

Camarero JA, Muir TW (1999) Native Chemical Ligation of Polypeptides. In: JF Coligan, BM Dunn, HL Ploegh, DW Speicher, PT Wingfield (eds): *Current Protocols in Protein Science.* John Wiley & Sons, Inc., 18.4.1–18.4.21.

Noren CJ, Wang J, Perler FB (2000) Dissecting the Chemistry of Protein Splicing and its Applications. *Angew Chem Int Ed* 39: 450–466.

Paulus H (2001) Inteins as Enzymes. *Bioorg Chem* 29: 119–129.

Perler FB, Adam E (2000) Protein splicing and its applications. *Curr Opin Biotechnol* 11: 377–383.

InBase, a curated on-line intein database (<http://www.neb.com/neb/inteins.html>) also provides an up-to-date, annotated intein bibliography. Perler FB (2002) InBase: The Intein Database. *Nucleic Acids Res* 30: 383–384.

Xu MQ, Paulus H, Chong S (2000) Fusions to self-splicing inteins for protein purification. *Methods Enzymol* 326: 376–418.

References

1   Gimble FS, Thorner J (1992) Homing of a DNA endonuclease gene by meiotic gene conversion in *Saccharomyces cerevisiae. Nature* 357: 301–306

2   Xu M, Southworth MW, Mersha FB et al. (1993) *In vitro* protein splicing of purified precursor and the identification of a branched intermediate. *Cell* 75: 1371 1377

3   Chong S, Mersha FB, Comb DG et al. (1997) Single-column purification of free recombinant proteins using a self-cleavable affinity tag derived from a protein splicing element. *Gene* 192: 271–281

4   Chong S, Montello GE, Zhang A et al. (1998) Utilizing the C-terminal cleavage activity of a protein splicing element to

purify recombinant proteins in a single chromatographic step. *Nucleic Acids Res.* 26: 5109–5115

5   Hoang TT, Ma Y, Stern RJ et al. (1999) Construction and use of low-copy number T7 expression vectors for purification of problem proteins: purification of *Mycobacterium tuberculosis* RmlD and *Pseudomonas aeruginosa* LasI and RhlI proteins, and functional analysis of purified RhlI. *Gene* 237: 361–371

6   Southworth MW, Amaya K, Evans TC et al. (1999) Purification of proteins fused to either the amino or carboxy terminus of the *Mycobacterium xenopi* gyrase A intein. *BioTechniques* 27: 110–120

7   Evans TC, Jr., Benner J, Xu MQ (1999) The *in vitro* ligation of bacterially expressed proteins using an intein from *Methanobacterium thermoautotrophicum*. *J Biol Chem* 274: 3923–3926

8   Mathys S, Evans TC, Chute IC et al. (1999) Characterization of a self-splicing mini-intein and its conversion into autocatalytic N- and C-terminal cleavage elements: facile production of protein building blocks for protein ligation. *Gene* 231: 1–13

9   Wood DW, Wu W, Belfort G et al. (1999) A genetic system yields self-cleaving inteins for bioseparations. *Nat Biotechnol* 17: 889–792

10 Wood DW, Derbyshire V, Wu W et al. (2000) Optimized single-step affinity purification with a self-cleaving intein applied to human acidic fibroblast growth factor. *Biotechnol Prog* 16: 1055–1063

11 Paul R, Bosch FU, Schafer KP (2001) Overexpression and purification of *helicobacter pylori* flavodoxin and induction of a specific antiserum in rabbits. *Protein Expr Purif* 22: 399–405

12 Thomson CA, Ananthanarayanan VS (2001) A method for expression and purification of soluble, active hsp47, a collagen-specific molecular chaperone. *Protein Expr Purif* 23: 8–13

13 Saiki K, Konishi K, Gomi T et al. (2001) Reconstitution and purification of cytolethal distending toxin of *Actinobacillus actinomycetemcomitans*. *Microbiol Immunol* 45: 497–506

14 Myscofski DM, Dutton EK, Cantor E et al. (2001) Cleavage and purification of intein fusion proteins using the *Streptococcus gordonii* spex system. *Prep Biochem Biotechnol* 31: 275–290

15 Pradhan S, Bacolla A, Wells RD et al. (1999) Recombinant Human DNA (Cytosine-5) Methyltransferase. I. expression, purification, and comparison of *de novo* and maintenance methylation. *J Biol Chem* 274: 33002–33010

16 Siebold C, Flukiger K, Beutler R et al. (2001) Carbohydrate transporters of the bacterial phosphoenolpyruvate: sugar phosphotransferase system (PTS). *FEBS Lett* 504: 104–111

17 Iwai H, Lingel A, Plueckthun A (2001) Cyclic green fluorescent protein produced in vivo using artificially split PI-PfuI intein from *Pyrococcus furiosus*. *J Biol Chem* 276: 16548–16554

18 Scott CP, Abel-Santos E, Jones AD et al. (2001) Structural requirements for the biosynthesis of backbone cyclic peptide libraries. *Chem Biol* 8: 801–815

19 Scott CP, Abel-Santos E, Wall M et al. (1999) Production of cyclic peptides and proteins *in vivo*. *Proc Natl Acad Sci USA* 96: 13638–13643

20 Camarero JA, Muir TW (1999) Biosynthesis of a head-to tail cyclized protein with improved Biological Activity. *J Amer Chem Soc* 121: 5597–5598

21 Zhang A, Gonzalez SM, Cantor EJ et al. (2001) Construction of a mini-intein fusion system to allow both direct monitoring of soluble protein expression and rapid purification of target proteins. *Gene* 275: 241–252

22 Dawson PE, Muir TW, Clark-Lewis I et al. (1994) Synthesis of proteins by native chemical ligation. *Science* 266: 776–779

23 Muir TW, Sondhi D, Cole PA (1998) Expressed protein ligation: A general method for protein engineering. *Proc Natl Acad Sci U S A* 95: 6705–6710

24 Kinsland C, Taylor SV, Kelleher NL et al. (1998) Overexpression of recombinant proteins with a C-terminal thiocarboxylate: implications for protein semisynth-

esis and thiamin biosynthesis. *Protein Sci* 7: 1839–1842

25 Roy RS, Allen O, Walsh CT (1999) Expressed protein ligation to probe regiospecificity of heterocyclization in the peptide antibiotic microcin B17. *Chem Biol* 6: 789–799

26 Welker E, Scheraga HA (1999) Use of benzyl mercaptan for direct preparation of long polypeptide benzylthio esters as substrates of subtiligase. *Biochem Biophys Res Commun* 254: 147–151

27 Cottingham IR, Millar A, Emslie E et al. (2001) A method for the amidation of recombinant peptides expressed as intein fusion proteins in *Escherichia coli*. *Nat Biotechnol* 19: 974–977

28 Blaschke UK, Silberstein J, Muir TW (2000) Protein engineering by expressed protein ligation. *Methods Enzymol* 328: 478–496

29 Ayers B, Blaschke UK, Camarero JA et al. (1999) Introduction of unnatural amino acids into proteins using expressed protein ligation. *Biopolymers* 51: 343–954

30 Evans TC, Jr., Benner J, Xu MQ (1999) The cyclization and polymerization of bacterially expressed proteins using modified self-splicing inteins. *J Biol Chem* 274: 18359–18363

31 Tolbert TJ, Wong C-H (2000) Intein-mediated synthesis of proteins containing carbohydrates and other molecular probes. *J Am Chem Soc* 122: 5421–5428

32 Cotton GJ, Ayers B, Xu R et al. (1999) Insertion of a synthetic peptide into a recombinant protein framework: a protein biosensor. *J Am Chem Soc* 121: 1100–1101

33 Evans TC, Benner J, Xu M-Q (1998) Semisynthesis of cytotoxic proteins using a modified protein splicing element. *Protein Sci* 7: 2256–2264

34 Cowburn D, Muir TW (2001) Segmental isotopic labeling using expressed protein ligation. *Methods Enzymol* 339: 41–54

35 Wang D, Cole PA (2001) Protein tyrosine kinase csk-catalyzed phosphorylation of src containing unnatural tyrosine analogues. *J Am Chem Soc* 123: 8883–8886

$\textbf{15}$ Two-hybrid Assay in *Escherichia coli* K12

Gustavo Di Lallo, Patrizia Ghelardini and Luciano Paolozzi

Contents

1   Introduction

Many manifestations of living organisms are the result of protein activities that need the formation of the protein multimers themselves or of macromolecular complexes with other proteins. Numerous *in vivo* and *in vitro* approaches have been developed to facilitate the study of protein interactions [1–3].

The most popular is the widely used two-hybrid system that permits detection and selection ligands of practically any bait protein. Despite its success and its robustness, as a result of a decade of testing in hundreds of laboratories, this

Methods and Tools in Biosciences and Medicine
Prokaryotic Genomics, ed. by M. Blot
© 2003 Birkhäuser Verlag Basel/Switzerland

approach is far from being artifact-free. The reasons for these artifacts have in some cases been elucidated. A false negative may arise because the two partner proteins are not efficiently imported into the nucleus and/or are inefficiently folded. Alternatively, either partner may be titrated away by endogenous proteins. This problem becomes particularly severe when the proteins under study are yeast proteins. For this same reason, yeast proteins may act as a bridge between the two proteins under study, thereby causing an activation of the reporter gene even if the two query proteins do not interact directly. Furthermore, some proteins very often score as positive with several different baits; the reason for such a result is still not clear.

These considerations prompted the development of a number of alternative approaches that could overcome these limitations or at least complement the yeast two-hybrid method. Most of these are carried out in *E. coli* and, aside from the considerations above, the ability to transpose the power of the yeast two-hybrid method into *E. coli*, would be an advantage *per se* because of the ease with which this organism can be manipulated and also because of the efficiency of transformation that can be up to two logs higher than in yeast [4–7].

One of these systems exploits the structure and properties of the repressor, coded by the cI gene of bacteriophage λ, and is based on a variation of the two-hybrid approach [8, 9]. The active λ repressor is a homodimer formed by two cI molecules whose N-terminal portion is involved in DNA recognition and binding, whereas the carboxyl-terminal portion mediates the two subunits dimerization. Therefore, the N-terminal DNA-binding domain of the λ cI repressor dimerizes inefficiently unless a separate C-terminal dimerization domain is present. Thus, the fusion of the cI N-terminal domain with a heterologous protein will result in a functional λ repressor, which is able to bind strongly to its operator and therefore confer immunity to λ infection only when the heterologous protein is able to dimerize efficiently, i.e., only those proteins that mediate efficient dimerization of the chimera *in vivo* permit survival of the host cell upon infection with λ phage.

Although very useful, this system has strict limitations: the plasmid copy number is not easily controlled and it is exploitable only in *E. coli* K12 strains permissive to λ lysogenization. In addition, the λcI hybrid assay in its entirety is limited to *E. coli* and permits only the detection of homodimers formation. To overcome these limits, we developed a strategy that allows us to both evaluate the dimerization efficiency [10] and extend this kind of investigation to other bacterial genus, such as *Pseudomonas* and *Salmonella* (Di Lallo et al., unpublished results).

To extend this method to the study of protein heterodimerization, we constructed a chimeric operator formed by the two hemi-sites of P22 and 434 phage operators, respectively. This operator can be recognized and bound only by a hybrid repressor formed by two chimeric monomers, one with the N-terminal portion of phage 434 and the other with that of phage P22. The C-terminal domains of both are constituted by heterologous proteins (or protein domains) whose interaction ability is under investigation. Only those proteins that mediate efficient dimerization of the two chimeric repressor monomers *in vivo* permit the formation of a functional repressor able to bind the P22–434 hybrid operator and shut off the synthesis of a downstream reporter gene [11].

2   Materials and equipment

2.1   Equipment

Thermal cycler, electroporator, spectrophotometer.
Bacterial strains and plasmids: see Table 1 and Figure 1.

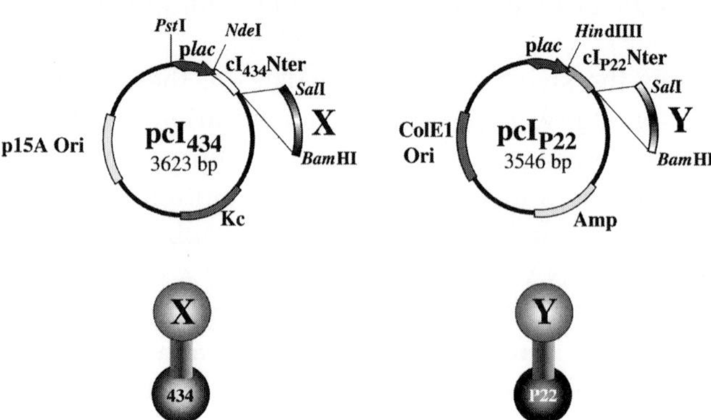

Figure 1   Schematic representation of the two plasmids pC434 and pC22 codifying the two chimeric repressors 434-X and P22-Y, respectively, where X and Y are the two proteins investigated.

Table 1   Bacterial strains and plasmids

Bacterial strains	Relevant genotype	Source
71/18	*supE thy* Δ(*lac-proAB*) F' [*proAB+ lacI^q lacZ*ΔM15]	[13]
R721	71/18 *glpT*::O-P$_{434/P22}$ *lacZ*	[11]
Plasmids		
pcI$_{434}$	pACYC177 derivative carrying the NH2 end of 434 repressor	[11]
pcI$_{P22}$	pC132 derivative carrying theNH2 end of P22 repressor	[11]

2.2 Media, solutions, and buffers

LB broth and agar for bacterial growth, SOC medium, for bacterial transformation:
- Bacto peptone   2 g
- Yeast extract   0.5 g
- NaCl 5M   0.2 ml
- KCl 1M   0.25 ml
- $MgCl_2$ 1M   1 ml
- $MgSO_4$ 1M   1 ml
- H_2O to 100 ml

Sterilize by autoclaving and add glucose, already sterilzed by filtration, to a final concentration of 20 mM.

Z-buffer, for β-galactosidase assay:
- $Na_2HPO_4.7H_2O$ 16.1 g (0.06M)
- $NaH_2PO_4.H_2O$ 5.5 g (0.04M)
- KCl0.75 g (0.01M)
- $MgSO_4.7H_2O$ 0.246 g (0.001M)
- β-mercaptoethanol   2.7 ml (0.05M)
- H_2O to 1000 ml

Do not autoclave, adjust pH to 7.0.

2.3 Chemicals

Ampicillin (50 µg/ml), kanamycin (30 µg/ml), chloramphenicol (34 µg/ml) (Sigma), *Sal*I, *Bam*HI restriction endonucleases and T4 DNA ligase (Biolabs), *Taq* DNA polymerase (Promega), *o*-nitrophenyl-β-D-galactopyranoside (ONPG) (Sigma), 4 mg/ml in water, Na_2CO_3 1M, and toluene.

3   Methods

The genes coding for the proteins of interest are cloned in the two vector plasmids pC434 and pC22, which contain the N-terminal domain of phage 434 and P22 repressor, respectively. The recombinat plasmids obtained are inserted by transformation into the *E. coli* strain R721 that harbors the 434-P22 chimeric operator upstream the *lacZ* gene. If the two proteins under investigation interact, a functional repressor is produced and the β-galactosidase produced by the bacterial strain is reduced (Fig. 2).

A Proteins or domains X and Y do not interact

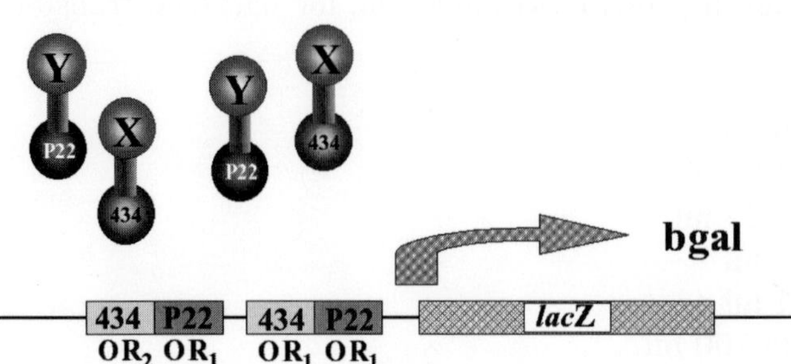

B Proteins or domains X and Y interact

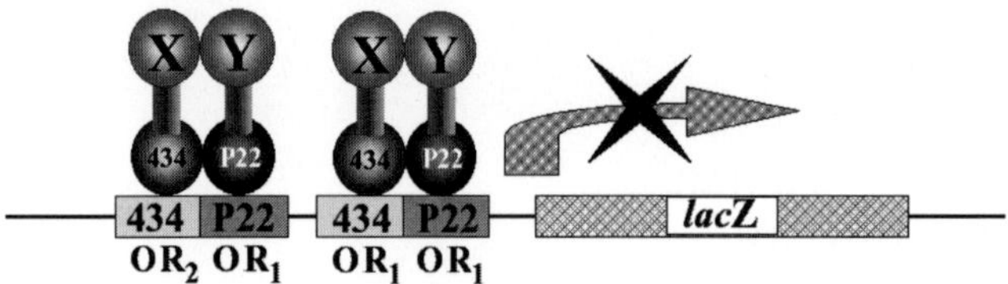

Figure 2   Schematic representation of the two-hybrid assay.
A) Proteins X and Y do not interact with each other, the funcional repressor is not formed. and the β-gal is normally expressed.
B) Proteins X and Y interact with each other, the functional repressor is formed, and the β-gal expression is repressed.

3.1   Cloning the genes of interest

Protocol 1   Primers design

The DNA of the gene of interest is obtained by PCR amplification using pairs of 28 bp synthetic oligonucleotides as reported below:
Forward: 5'GCGTCGACC plus specific sequence (19 nucleotides of the gene sequence starting from ATG).
Reverse: 5'CGGGATCC plus specific sequence (19 nucleotides of the gene sequence starting from the stop codon).

Protocol 2   PCR amplification

1. For each PCR reaction, combine:
   Genomic DNA (100 ng/µl)   1 µl
   10x PCR Buffer   5 µl
   dNTP (12.5 mM)   1 µl
   Primer 1 (50 µM)   1 µl
   Primer 2 (50 µM)   1 µl
   Taq Polymerase (5 U/µl)   1 µl
   Double distilled H₂O   40 µl

2. PCR amplification:

Step 1:

    Denaturation   94 °C x 50″
    Annealling   T1 x 50″
    Extension   72 °C x 1' for each Kb of DNA
This program is repeated for 5 cycles.

Step 2:

    Denaturation   94 °C x 50″
    Annealling   T2 x 50″
    Extension   72 °C x 1' for each Kb of DNA
This program is repeated for 25 cycles.

Step 3:

    Long extension   72 °C x 5'

3. Determination of T1 and T2:
Determine the oligonucleotide of the couple with the lower Tm.

4. Control the PCR product on agarose gel and recovery of the DNA band from the gel.

Note:

T1=Tm of the oligonucleotide specific sequence −5 °C
T2= Tm of the oligonucleotide specific sequence

Protocol 3   Cloning

1. Digest plasmids pC434 and pC22 with restriction endonucleases *Sal*I and *Bam*HI.
2. Digest the PCR DNA recovered from the gel with the same endonucleases for 24 h at 37 °C (these restriction sites are located in the primers used for the amplification).
3. After digestion, extract DNA by phenol/chloroform and precipitate with 2.5 volumes of absolute ethanol, centrifuge, and resuspend in TE pH 8.
4. For each plasmid, ligate about 100 ng of DNA vector with the equimolar amount of PCR product for 24 h at 16 °C in 10 µl of ligation buffer 1 X and 400 U of T4 DNA ligase (Biolabs).
5. Insert by electroporation (see Protocol 5) each recombinant plasmid in a suitable *E. coli* strain (such as 7118, XL1 blu, or αDH5), selecting for the appropriate antibiotic.
6. Control the recombinant plasmid by endonuclease digestion.

3.2 Assembling of the two-hybrid system

Protocol 4   Control of the *E. coli* R721 recipient strain

Isolate the strain on LB agar supplemented with 34 µg/ml of chloramphenicol and check the β-galactosidase production as described below (Protocols 8 and 9). The value of β-galactosidase should range between 2500 and 3000 Miller units.

Protocol 5   Preparation of *E. coli* R721 electro-competent cells

1. Inoculate one colony of R721 strain in 50 ml LB supplemented with 34 µg/ml of chloramphenicol and grow at 37 °C to $OD_{600} = 0.7$, centrifuge for 10 min at 4000 rpm at 4 °C.
2. Wash the pellet with the same volume of ice-cold double-distilled water, centrifuge, and wash again in the same conditions.
3. Centrifuge 10 min at 4000 rpm at 4 °C and wash the pellet with 10 ml ice-cold 10% glycerol.
4. Centrifuge again and resuspend the pellet in 300 µl ice-cold 10% glycerol.
5. Prepare aliquots of 50 µl in Eppendorf microfuge tubes. Store at –80 °C.

Protocol 6   Transformation of *E. coli* R721

1. To 3 aliquots of R721 competent cells add, respectively:
   10 ng of pC434-X plasmid DNA,
   10 ng of pC22-Y plasmid DNA,
   50 ng of pC434-X plasmid DNA plus 50 ng of pC22-Y plasmid DNA,
   where X and Y represent the two gene sequences cloned in the two vector plasmids.
2. Transfer each mix into 1 mm elecroporation cuvettes and electroprate at 1.8 Kvolt.
3. Recover each mix with 1 ml SOC and incubate at 37 °C for 1 h.
4. Plate on LB agar supplemented with:
   chloramphenicol (34 µg/ml) and kanamycin (30 µg/ml)
   chloramphenicol (34 µg/ml) and ampicillin (50 µg/ml)
   chloramphenicol (34 µg/ml) and kanamycin (30 µg/ml) plus ampicillin (50 µg/ml)
5. Purify the transformant clones by isolation on LB supplemented with the appropriate antibiotics.

3.3   Quantification of protein-protein interaction

Protocol 7   Growth conditions

1. Resuspend at least 5 colonies from the isolation plate in 0.5 ml LB.
2. Dilute 1:1000 in 10 ml LB supplemented with 10^{-4} M IPTG, and grow with aeration at 34 °C for about 5 h. At this time the OD_{600} should range between 0.3 and 0.4; otherwise, readjust the initial dilution.

Protocol 8   β-galactosidase assay [12]

The assays are performed in Eppendorf microfuge tubes.
1. To an aliquot of 300 μl from the culture add 700 μl of Z-buffer and 25 μl of Toluene.
2. Vortex for 30″.
3. Evaporate Toluene for 20–30 min.
4. Recover 700 μl from the bottom of the tube and transfer them to another tube.

To start the reaction:
5. Transfer in a water bath at 30 °C.
6. Add 140 μl ONPG (4 mg/ml).
7. Start the chronometer.
8. Stop the reaction by adding 300 μl Na_2CO_3 1M after sufficient yellow color has developed.

Stop the chronometer and record the reaction time.
9. Centrifuge the samples 2 min at 14,000 rpm.
10 Transfer the supernatant into spectrophotometer cuvettes and read the OD at 420 nm against a blank constituted by:
   185 μl LB
   H_2O to 1 ml

Protocol 9   Determination of β-galactosidase units and valuation of protein protein interaction

Miller units of β-galactosidase

$$= \frac{OD_{420} \times 1000}{OD_{600} \times V \times T}$$

where V = volume of the sample = 0.3 ml; T = reaction time in minutes.

Make the ratio between the β-galactosidase units produced by the R721 strain containing both plasmids carrying the two proteins under investigation and that produced by R721 without plasmids. For interacting proteins, this ratio should be less than 0.5.

4 Troubleshooting

This two-hybrid assay has been applied with success to a high number of proteins. An intrinsic limitation of this system could be the uncertainty of the results for weak or transient interactions (residual β-gal activity between 50% and 70%). In these cases, an accurate statistical evaluation of the results is necessary.

5 Applications

This assay can be used to test ligands, both prokaryotic and eukaryoric, with a wide range of size (from less than 100 to more than 800 amino acids) and binding affinities. No limitation has been observed using this system to test the interaction between membrane-localized proteins (Di Lallo et al., unpublished results).

6 Remarks and conclusions

At the moment, because of its feasibility, reproducibility, and low cost, the lambdoid repressor dimerization strategy is a convenient method to test *in vivo* protein protein interactions and/or to confirm those already suggested by other genetical or biochemical studies.

Acknowledgements

We thank M. Lo Ponte for revision of the manuscript.

This work was supported by funds from the Ministerio dell'Istruzione, dell'Università della Ricerca (Italy).

References

1   Fields S, Song O (1989) A novel genetic system to detect protein-protein interactions. *Nature* 340: 245–246

2   Hanes J, Pluckthun A (1999) *In vitro* selection methods for screening of peptide and protein libraries. *Curr Top Microbiol Immunol* 243: 107–122

3   Vidal M, Legrain P (1999) Yeast forward and reverse 'n'-hybrid systems. *Nucleic Acids Res* 27: 919–929

4   Dove SL, Joung JK, Hochschild A (1997) Activation of prokaryotic transcription through arbitrary protein-protein contacts. *Nature* 386: 627–630

5   Karimova G, Pidoux J, Ullmann A, Ladant D (1998) A bacterial two-hybrid system based on a reconstituted signal transduction pathway. *Proc Natl Acad Sci USA* 95: 5752–5756

6   Kornacker MG, Remsburg B, Menzel R (1998) Gene activation by the AraC protein can be inhibited by DNA looping between AraC and a LexA repressor that interacts with AraC: possible applications as a two-hybrid system. *Mol Microbiol* 30: 615–624

7   Karimova G, Ullmann A, Ladant D (2000) A bacterial two-hybrid system that exploits a cAMP signaling cascade in Escherichia coli. *Methods Enzymol* 328: 59–73

8   Hu JC, O'Shea EK, Kim P, Sauer RT (1990) Sequence requirements for coiled-coils: analysis with l repressor-GCN4 leucine zipper fusions. *Science* 250: 1400–1403

9   Longo F, Marchetti MA, Castagnoli L, Battaglia PA, Gigliani F (1995) A novel approach to protein-protein interaction: complex formation between the p53 tumor suppressor and the HIV Tat proteins. *Biochem Biophys Res Commun* 206: 326–334

10 Di Lallo G, Ghelardini P, Paolozzi L (1999) Two-hybrid assay: construction of an *Escherichia coli* system to quantify homodimerization ability *in vivo*. *Microbiol* 145: 1485–1490

11 Di Lallo G, Castagnoli L, Ghelardini P, Paolozzi L (2001) A two-hybrid system based on chimeric operator recognition for studying protein homo/heterodimerization in *Escherichia coli*. *Microbiol* 147: 651–1656

12 Miller JH (1972) *Experiments in molecular genetics*. Cold Spring Harbor Laboratory Press, Cold Spring Harbor, NY

13 Dente L, Cesareni G, Cortese R (1983) pEMBL: a new family of single stranded plasmids. *Nucleic Acid Res* 11: 1645–1655

Guide to Protocols

Index